W9-DDJ-178

Creating Child Friendly Cities

Creating Child Friendly Cities assesses the extent to which the physical and social make-up of Western cities accommodates and nourishes the needs of children and young people. Featuring contributions in the areas of planning, design, social policy, transport and housing, the book outlines strengths and deficiencies in the processes that govern urban development and change from the perspective of children and young people.

The book responds to continuing interest in the welfare of children in cities and resonates with growing international concerns about the health and well being of young people in Western countries. It presents a critical analysis of the issues facing those dealing with children's needs in the urban environment – such as the 'obesity epidemic', transportation issues and the changing role of parenting. This is essential reading for professional planners and policy makers seeking answers to the challenges of creating cities that work for and include children.

Brendan Gleeson is Director of the Urban Research Program at Griffith University, Brisbane. Before joining Griffith he was Deputy Director of the Urban Frontiers Program at the University of Western Sydney. His research interests include urban planning and governance, urban social policy, disability studies, and environmental theory and policy.

Neil Sipe is Head of the Environmental Planning School at Griffith University and is an experienced urban researcher who has worked in North America and Australia. He has an extensive teaching record in the field of transport planning and in recent research has been the first Australian scholar to propose methods for defining and mapping transport exclusion.

This is for children in cities,
especially for Julian, Alison, Caroline
and Catherine

Creating Child Friendly Cities

Reinstating kids in the city

Edited by Brendan Gleeson and Neil Sipe

Routledge
Taylor & Francis Group
LONDON AND NEW YORK

First published 2006
by Routledge
2 Park Square, Milton Park, Abingdon, Oxon OX14 4RN

Simultaneously published in the USA and Canada
by Routledge
270 Madison Ave, New York, NY 10016

Routledge is an imprint of the Taylor & Francis Group, an informa business

Transferred to Digital Printing 2007

Typeset in Galliard and Helvetica by
Florence Production Ltd, Stoodleigh, Devon
Printed and bound in Great Britain by
Antony Rowe Ltd, Chippenham, Wiltshire

British Library Cataloguing in Publication Data
A catalogue record for this book is available from the British Library

Library of Congress Cataloging in Publication Data
Creating child friendly cities: reinstating kids in the city/edited by
Brendan Gleeson and Neil Sipe.
p. cm.
Includes bibliographical references and index.
1. City children. 2. Cities and towns. 3. City planning.
4. Child welfare. 5. Children and the environment. I. Gleeson,
Brendan, 1964– II. Sipe, Neil G.
HT206.C74 2006
307.1'216–dc22 2006004300

ISBN10: 0–415–39160–1 (hbk)
ISBN10: 0–203–08717–8 (ebk)

ISBN13: 978–0–415–39160–3 (hbk)
ISBN13: 978–0–203–08717–6 (ebk)

Contents

Illustrations

Figures

Tables

Contributors

Nick Buchanan is a Transport Planner with the consultancy GHD Pty Ltd in Sydney. Nick is involved in a variety of transport-related projects, from improving travel choices at the local level, up to developing strategic national level transport corridors between Australian cities. Prior to this, he was a Senior Research Assistant in the Urban Research Program at Griffith University, Brisbane. His interests focus on how urban form and other influences affect individual travel patterns.

Damian Collins is a Post-doctoral Fellow in the School of Geography and Environmental Science at the University of Auckland, New Zealand. He is particularly interested in public schools as social and legal spaces, and potential sites of health promotion. Together with Robin Kearns, he has written several articles on the health, well-being and legal status of children in New Zealand.

Jago Dodson is a Research Fellow in the Urban Research Program at Griffith University, Brisbane. Jago's published research covers urban governance, planning, housing and transport. Jago is presently completing a book on public housing policy in Australia and New Zealand and has ongoing projects examining various aspects of urban development, including transport disadvantage, transport stress, accessibility and the impact of urban environments on health as well as the role of corruption in the planning process.

Claire Freeman is currently Director of the Planning Programme at the University of Otago, Dunedin, New Zealand. Claire came to Otago via lecturing posts at Massey University, Leeds Metropolitan University and the University of the North West in South Africa, and has been a planner for the Urban Wildlife Trust in Birmingham, England. Her research interests are in planning with children, planning for nature and community planning.

Brendan Gleeson is Director of the Urban Research Program at Griffith University, Brisbane. Before joining Griffith in March 2003, Professor Gleeson was Deputy Director of the Urban Frontiers Program, University of Western Sydney. His research interests include urban planning and governance, urban social policy, disability studies, and environmental theory and policy.

He is co-author (with Nicholas Low) of *Justice, Society and Nature: An Exploration of Political Ecology* (Routledge 1998). This book received the prestigious Harold and Margaret Sprout Award in 1999 from the International Studies Association. He has also co-edited three books with Nicholas Low on aspects of urban and environmental policy. Professor Gleeson's urban social policy interests were reflected in his 1999 book, *Geographies of Disability.* In 2001, his book (with N. P. Low) *Australian Urban Planning: New Challenges, New Agendas* received the Royal Australian Planning Institute's National Award for Planning Scholarship Excellence.

Kurt Iveson is a Lecturer in Urban Geography at the University of Sydney. Kurt is particulary interested in the nature of conflicts over the urban public realm. This interest emerged from his previous experiences as a policy officer for the Youth Policy and Action Association of New South Wales and the Australian Youth Policy and Action Coalition, and as a board member of the Youth Coalition of the Australian Capital Territory. His book *Publics and the City* (Blackwell) will be published in 2006.

Robin Kearns is Professor of Geography in the School of Geography and Environmental Science at the University of Auckland, New Zealand. He has research interests in public health, cultural geography and qualitative research methods. He is co-author (with Wilbert Gesler) of *Culture/Place/Health* (Routledge 2002). He has been personally and professionally involved in the development of walking school buses in Auckland since 1999.

Karen Malone is Asia-Pacific Director of the global UNESCO Growing Up In Cities (GUIC) project. GUIC focuses on children and youth evaluating the quality of their local environments, and the project recently won the internationally prestigious Environmental Design Research Association most outstanding research award. Karen Malone is located in the Centre of Interdisciplinary Youth Research at the Faculty of Education, University of Wollongong, New South Wales. She has published widely in the areas of children and youth environments and is currently completing a book on the changing nature of transitions from childhood to youth in the twenty-first century entitled *From Hopscotch to Hip Hop.*

Kylie Rolley is a fourth-year student of Environmental Planning at Griffith University, Brisbane, and is currently undertaking her honours thesis looking at social disadvantage in peri-urban south-east Queensland. Kylie also works at the Urban Research Program where she is employed as a Research Assistant.

Neil Sipe is Head of the Environmental Planning School at Griffith University, Brisbane, and is an experienced urban researcher who has worked in North America and Australia. He has an extensive teaching record in the field of transport planning and in recent research has been the first Australian scholar to propose methods for defining and mapping transport exclusion. He has a strong track record in empirical research that links issues of spatial access and socio-economic equity in urban contexts, both in the US and in Australia.

Paul Tranter is a Senior Lecturer in Geography in the School of Physical, Environmental and Mathematical Sciences at the University of New South Wales at ADFA (the Australian Defence Force Academy) in Canberra. Here he lectures in social geography and transport geography. His research interests include the themes of child friendly environments and sustainable cities, the public health impacts of motor sport and the promotion of active transport through the concept of 'effective speed'. His research on children includes the examination of children's freedom to explore their own neighbourhoods in Australia and New Zealand, as well as children's freedom to play within their own school grounds.

Prue Walsh is an Early Childhood Educator who has developed specialized expertise in play environment planning and safety. Over the last 20 years she has been involved with on-the-ground planning of over 1,800 early childhood centres, schools and public community facilities. She has brought to this role an in-depth understanding of how and why children play and the type and design of facilities needed to draw out children's active engagement in play. This work has covered all states in Australia, and also Europe, Asia and America. Currently she is a member of the Australian Playground Standards committee. In 1991 she was awarded a Churchill Fellowship for her work in this area.

Acknowledgements

This book was conceived during a symposium, *Creating Child Friendly Cities*, conducted in Brisbane in October 2004. The symposium was generously supported by Delfin Lend Lease Corporation, and drew scholars, activists and professionals from a wide range of sectors in Australia and New Zealand. The editors thank all those who participated, including presenters not represented in this volume, for helping to create an inspiring and insightful meeting. The symposium underlined the need for a multidisciplinary, cross-sectoral approach to the creation of improved urban environments for children.

The editors are deeply grateful to Ms Kylie Rolley and Ms Gillian Warry, both at Griffith University, Brisbane, for their assistance in the preparation of this manuscript. Kylie's help throughout 2005 was especially invaluable.

Chapter 1

Reinstating kids in the city

Brendan Gleeson and Neil Sipe

Recent years have seen a rapidly growing public debate about the welfare of children in Australia and New Zealand. These have resonated with wider international discussions that are responding to new professional concerns about the health and well-being of young people in Western countries. The Australian popular media is awash with reportage on issues such as childhood obesity, psychic stress in children and young people, the neglected transport needs of the young, and concern about child exposure to abuse and other forms of harm.

Much of this reportage has an urban cast, suggesting increased popular recognition that the health and well-being of children have direct corollaries in dimensions of urban development. For example, a widely read and cited 2004 essay in the *Sydney Morning Herald* made explicit links between the epidemic of childhood obesity and new forms of urban development that consigned children to sedentary lives. The wealthier areas of Australia's cities were held to contain a 'bubble wrap generation' – 'pampered prisoners' whose possibilities for recreation and self-expression were limited by poor residential design and by high levels of parental anxiety and control (Cadzow 2004: 18–21).

The contemporary debates on children's well-being in Western, especially English-speaking, countries have two defining qualities. First, they mark a resurgence of concern for children in professional, political and popular quarters after a period of declining apparent interest in the well-being of young people. Arguably, other population groups have claimed the centre stage of public debate since the 1970s. They include the swelling legions of older people in the West's ageing societies, new migrant populations that have aroused social interest (and, lamentably, occasional social antipathy), and the social groups that from the 1960s sought liberation from older repressive moral and institutional orders, including women, gay people and disabled people.

The second defining quality of the new debates on children is their increasing multidisciplinarity, reflecting new professional and scientific recognition of the interdependencies between the different dimensions of children's health and well-being. Increasingly, the traditionally specialized understandings of children's health are opening out to recognize the broad range of factors in the everyday environment that influence the physical and mental condition of children. This is a key point of implication in the new debates for urban scholars and policy makers, whose understandings about the creation and experience of social space are increasingly sought by public health experts, child psychologists and child educationalists. Childhood experts show increasing recognition of, and interest in, the ways in which built environments both reflect and condition the key environmental and behavioural dynamics that shape the well-being of children.

In Australia, these two qualities of the new debates on children were manifest in the 2005 publication *Children of the Lucky Country?* (Stanley, Richardson and Prior). The book has been a bestseller in a subject area not previously known for mass sales, and was marketed and sold in the main variety stores that populate suburban shopping malls. The book was co-authored by a widely esteemed epidemiologist,[1] an economist and a psychologist, reflecting the new possibilities, indeed imperatives, for interdisciplinary discussion of children's issues. The book's sub-title – *How Australian Society has Turned its Back on Children and Why Children Matter* – supports the view that children have been suffering from impoverished popular and political attention in recent decades, with deleterious consequences for their well-being. The three authors, none of whom are recognized as urban scholars, go to considerable lengths in the book to explore how the qualities of built environments help to determine the life chances of children. These explorations are limited and preliminary and invite a broader, more thoroughgoing engagement between urbanists and the various professions with dedicated interests in children.

Children of the Lucky Country? presents a disturbing account of how children's interests have been sidelined by the rise of neo-liberalism and the consequent growth of materialism and individualism. The authors also argue that politicians, professions and institutions have overreacted to major demographic shifts – notably, the growth of smaller households, population ageing – and have assumed that children are no longer central priorities for politics and policy. The book decries the marginalization of children, and demonstrates the potential for this shift to weaken the social structure of Australia. The authors maintain that a society that neglects children is not concerned with its future in general, and is vulnerable thereby to social and ecological dysfunction. A society that places children's needs at its centre must, of necessity, always look to the future and make provision for it. Children remind us of the meaning and the urgency of the imperative for inter-generational equity, which goes to the core of sustainability. A sustainable society, as *Children of the Lucky Country?* points out, places children at its centre.

We share this commitment, which underlines the key hope of this book: that children will be reinstated to the centre of debate and analysis of urban conditions. This is a project that is necessarily beyond the scope of this book, which aims to show some ways in which this reinstatement might occur. Our objectives in pursuit of this aim are to show how children's needs can be identified and addressed at a variety of policy and conceptual scales and across a range of professional interests, all linked by the framework provided by urban environments. Our definition of children is encompassing, spanning the entire non-adult range, from infants to youth. Most of the studies in this volume cover this age span, though one chapter specifically focuses on older children, described both as youths and young people.

Importantly, *Creating Child Friendly Cities* conceives the city as more than simply a design problem (or opportunity) for children. Its view of the urban recognizes the centrality of institutional structures, professional understandings and social practices to the constitution of city life and, thereby, to child well-being. Before going on to foreshadow how the book will address its aim and objectives, it is worthwhile pausing to consider why we feel the need, with other commentators, to reinstate kids in the city. In other words, by what circumstances did they get swept aside from the mainstreams of social enquiry and popular debate?

Lost in space: the rise and fall of kids in the city

Modern conceptions of childhood and child well-being emerged in the wake of the Industrial Revolution and were crystallized during the class struggles, environmental changes and demographic shifts that were unleashed by modernization. Children were the first 'poor creatures' that mid and late Victorian reform movements sought to rescue from the hellfire cities that emerged in the wake of first-wave industrialism. On the question of child well-being, there was a decided unity of understanding and purpose – a corollary of what we might now term interdisciplinarity – among the various sanitary, labour and housing reformers who sought to check the course of a raw capitalism that was careering towards a social and ecological precipice. Town planning emerged as part of a wave of social improvements that sought, among other things, to safeguard children. Whether consciously alert to their deeper purpose or not, the Victorian improvers seemed at least instinctually aware that by rescuing the 'vulnerable' (children, then women, then the proletariat) from the maw of industrial capital, the reform project was in fact rescuing capitalism from an increasingly apparent will for self-destruction. By securing the material welfare of the vulnerable, reform guaranteed a future for capitalism and diverted politics away from the revolutionary cataclysm, which some by the late nineteenth century believed was inevitable.

Reform was overtaken by the currents of class struggle diverted from the prospect of revolution and conflict, and a new improvement project produced the welfare state in the twentieth century. Modernization continued with a political licence that stipulated the need for constant material improvement, particularly for the working class and vulnerable groups. Children were submerged in this social compact, but remained at its centre. Their sheer demographic significance in a time of rapid population increase, especially after the Second World War, ensured a political and social centrality for children. In the everyday practice and thinking of the professions that created and recreated cities, children were an assumed central consideration – indeed, so deeply assumed that some commentators became concerned that urban professional practice, though relentless in its focus on children, was also thoughtless about their particular needs. Were they another artefact of modernity, mass produced and mass provided for?

Thus, in North America and Europe a lively literature emerged during the 1960s that attempted to give more explicit thought to the links between urban development and children's welfare. Colin Ward's *The Child in the City* (1978) distilled this complaint with industrial modernism and proposed an urbanism that was much more conscious of children's diverse needs, including their abiding preference for secure home worlds over broad cityscapes. The growing critical focus on children among urban commentators was stimulated by the establishment of a ten-year programme in 1968, called *Growing Up in Cities*, coordinated by the United Nations Educational, Scientific and Cultural Organization (UNESCO). Much of this discussion focused on highly particular questions, such as how aspects of child psychology were influenced by environmental conditions (e.g. Gump 1975) or narrow concerns with the physical design of child play areas (e.g. Lady Allen of Hurtwood 1968). Overall, the ambition of these projects and commentaries was not so much to re-centre children socially as to urge greater institutional awareness of their unique and sensitive qualities (e.g. Lynch 1977). Children were still at the core of the modernization project, but institutions were behaving zombie-like, providing thoughtlessly for their assumed, not considered, needs.

During the 1980s research into children's issues mostly continued with the themes established during the 1970s. That is, it focused on how the physical environment impacted on the social and mental development of children (e.g. Homel and Burns 1985). Some attempts were made to understand the environment from a child's perspective and incorporate these ideas into policy, but by and large the emphasis remained on children's development and how that is shaped by the physical environment. National and international debates on children and cities had quietened by the 1990s.

As noted earlier, recent years have witnessed renewed interest in public and professional discussion of urban children's issues in English-speaking countries. Specifically, children's physical health has emerged as an area of sharp concern with the recognition that levels of physical fitness among urban children have been declining, most notably in Western countries. Scholarly research and popular interest in children's health has continued into the twenty-first century, focusing particularly on the incidence of childhood obesity and the associated decrease in children's physical activity. Responding to concern about childhood obesity, a growing range of studies has examined the links between children's physical activity patterns and built environment form. Other investigations have pointed to an alarming rise in mental health disorders among children in countries such as the United States, the United Kingdom and Australia.

The sense of urgency that appears to characterize new assessments of children's well-being seems charged by the view that children have been downgraded or even swept aside as a political concern and as an institutional priority. There are several potential dimensions to this claim. First, a demographic shift has been under way in developed countries towards smaller households and fewer children. Population ageing has become a key political and institutional concern, arguably to the exclusion of children's issues. Second, the rise of neo-liberalism, especially in English-speaking countries has been marked by the ascendancy of economic priorities over social priorities in political and institutional realms. In this context, econocratic thinking fixes on Economic Wo/Man – the consumer/taxpayer. Children do not make the balance sheet. (The influence of neo-liberal political and institutional practice on the well-being of children is an important theme in this volume.)

Importantly, the new debates that are attempting to refocus political and institutional attention on children frequently resonate with strident criticism of neo-liberalism and its socio-political consequences, such as increased social and residential segregation, environmental depletion and injury, and strengthened consumerism and materialism. In Australia, *Children of the Lucky Country?* (Stanley, Richardson and Prior 2005) largely attributes the rising indications of morbidity among young Australians to the growth of social inequality and the increased competitive pressures on families and individuals. The authors implicate urban transformations in the decline of child welfare, arguing that economic change has created a geography of winner and loser neighbourhoods in the cities. Children sorted or born into the new urban poverty concentrations face relatively poor life prospects.

Just as challengingly for the neo-liberal model, an accumulation of scientific evidence suggests that growing material wealth poses very real physical and psychological risks for children. Luthar's (2003) survey of evidence points to the heavy psychological costs that American children are paying for 'the culture of affluence' that has been contrived by contemporary neo-liberalism. This criticism, of course, is not confined to neo-liberalism and questions a deep assumption of modernization generally: that rising and generalizing affluence would drive a mass improvement in children's well-being. The criticism was echoed in a popular book in the United States entitled *Home-Alone America: The Hidden Toll of Day Care, Behavioral Drugs, and Other Parent Substitutes* (2004). The book's author, Mary Eberstadt, reported skyrocketing rates of depression, anxiety and behavioural disorders among children and teenagers in middle-class and wealthier families. Commentators have taken issue with Eberstadt's causal assessment – especially her critique of day care and working motherhood – but there is rising agreement among childhood experts in the United States that many middle-class children are suffering from parental deprivation. The steadily accumulating evidence reviewed and discussed by these works points to a much more complex, even fraught, relationship between household wealth and child well-being.

This book emerges at a time when debates about children echo those of the Victorian reform period. Some may scoff at this comparison. But in so doing they ignore the stridency of concern emanating from child health experts who report an alarming decline in the well-being of children assessed against a variety of mental and physical health indicators. These claims from experts, not advocates or activists, suggest that many Western children are imperilled by the socio-economic and environmental pressures bearing down on them and by the institutional disregard for their worsening circumstances. The authors of *Children of the Lucky Country?* state for Australia: '[t]he present generation of children may be the first in the history of the world to have lower life expectancy than their parents' (Stanley, Richardson and Prior 2005: 52). 'Child rescue' appears to be back on the agenda, as in the nineteenth century, again with a strong emphasis on fundamental health issues and their basis in urban conditions.

But against this background of rising professional alarm, we can identify strong resources for hope. Modernization has hardly been an unmitigated failure. Our societies possess the wealth, the science and, critically, the reflexive sensibility that is needed to create the conditions in which children can flourish. Additionally, the new interdisciplinary alliances and understandings that are emerging in response to the renewed concern for children are surely the bases for the powerful institutional and professional interventions needed to create these conditions.

Urban analysis is essential to this improved scientific understanding of children's contemporary problems and needs. Urban environments are where the vast majority of people in Western countries reside and are the principal context within which we must provide flourishing conditions for children. But they are more than mere context: cities and suburbs are dynamic, even fluid, social spaces where constant transformation acts independently to shape the communities that inhabit them. This recognition is seeping through to the non-urban professions – in health, community development, education, recreation – who look increasingly to urban analysis for enhanced understanding of how complex environments influence the well-being of children.

This volume responds to this recognition. It is firmly, if broadly, rooted in urban analysis: its chapters draw from expertise in urban planning, social geography, child play design, transport planning and education. And yet it is framed in a way that seeks to draw into its urban analysis other professional interests and knowledge. Several chapters, for example, relate their analyses of changing urban conditions to shifts in health outcomes for children. Of course, this is not the 'full story' on the child friendly city: it is a contribution to the broader project of reinstating children at the centre of community interests and institutional priorities, and it underlines the importance of urban analysis to this project. There is a message here for urban studies – especially geography and urban planning – which have not in recent times produced a strong stream of inquiry into children. This work is intended to contribute to a reawakening in urban studies of the potential for child-focused scholarship to improve general understandings of city formation, with an emphasis here on governance and professional practice.

We follow and have benefited from the studies that have attempted urban inquiries into children's well-being, such as Christensen's and O'Brien's 2002 collection, *Children in the City*. Our ambition differs somewhat from that study, which emphasized children's experience of urban conditions, a critical premise for effective institutional practice in the child friendly city. Our volume examines children in the city from a different, though complementary, perspective. It explores the different institutional and spatial scales at which cities can be conceived from the perspective of children, and examines the domains of policy and professional practice that shape urban conditions for children.

The structure of the book

The book is structured in three parts. It seeks three ways of examining the institutional and professional forces that shape cities for children. We cannot in one volume examine all or even most of these forces, so our survey takes three 'core samples' of the institutional and professional terrain. Part One examines the different politico-spatial scales at which the urban welfare of children can be conceived and studied. The scales examined here are the international including the United Nations (UN), the national (focusing on Australia) and the metropolitan (Perth and Sydney). Karen Malone (Chapter 2) addresses the United Nations Child Friendly City global initiative. She reminds us that there is a global context for the discussion of children's urban well-being in the form of a *Convention on the Rights of the Child* (CRC) and an array of programmes and plans formulated by the United Nations and its agencies. Brendan Gleeson (Chapter 3) explores some of the implications of urban structural change in Australia for children at both ends of the wealth scale. He charts the emergence of 'toxic cities': urban areas that fail to nurture the young and that increasingly threaten them with physical and mental harm. Gleeson agrees with Malone's statement that children are 'canaries in the mines' and that a national society that ignores them is casting aside its future. Kurt Iveson (Chapter 4) provides a critical review of official approaches to young people's use of public space in two of Australia's largest cities, Perth and Sydney. He shows that in both metropolitan examples, campaigns to secure the city for a community said to be under threat from 'unruly youth' have produced a range of new measures designed to exclude young people and criminalize 'irresponsible' parenting. The child friendly city at this scale will attempt not to control and pacify its 'unruly young', but respect their right to participate in a broader democratic debate about what constitutes the good city and good behaviour.

Part Two considers two important policy domains that contain strong professional groupings that, in different ways, have the potential to craft urban environments in child friendly ways. These domains are planning and health. Both share a common heritage in the era of reform that improved the Victorian city, and both have been guided by professions with a strong commitment to human improvement. In the twentieth century they drifted apart, but they appear to be finding new grounds for collaboration. This is partly driven by recent recognition that the declining health of many, if not most, Western children has a uniquely urban genesis. Claire Freeman (Chapter 5) notes the proliferation of initiatives directed at children and young people's participation in local government in the United Kingdom and New Zealand. Planners, she believes, are increasingly aware of children's and young people's social and environmental needs and rights. However, her assessment is that progress towards more child-focused planning has been limited and the initiatives ad hoc. Freeman outlines a framework for a more thoroughgoing engagement of planning, especially at the local level, with children's issues. Neil Sipe, Nick Buchanan and Jago Dodson (Chapter 6) review the connections between children's health and the nature of the urban environment. They concentrate on those discussions and research investigations in Western countries that have linked aspects of urban form with the mental and physical health of children. The debates have evolved in recent decades from relatively self-contained disciplinary discussions (public health, psychology, physical medicine, urban planning) towards an increasingly connected set of discussions that recognize the many interdependent relationships that link built environments with health outcomes. As part of this, health professionals have become more aware of the need for scientific engagement with urban professions and urban scholars.

Part Three takes us to specific spheres of policy or programmatic action within the city. We choose three urban 'theatres' that are especially important to the well-being of children in the contemporary city. Robin Kearns and Damian Collins (Chapter 7) consider the case of the walking school bus, an increasingly popular response to the dual concerns of traffic congestion and child health and safety in the intensifying city. They review the growth of walking school buses in Auckland, New Zealand, noting the practical steps and strategic alliances that have ensured the programme's success. Walking school buses promote child health, but they provide no simple panacea for the many problems children face in urban areas, including pedestrian safety and the harm that derives from traffic congestion. Only the more thoroughgoing child-centred policy focus, such as that urged by Freeman for planning, can address the structural factors that produce dangerous spheres for children. These factors include motor car dependence, economic inequality and poor public health.

Next Paul Tranter (Chapter 8) unpacks the notion of the 'social trap', a highly problematical sphere of *inaction* for children in the contemporary city. Social traps occur where parents feel compelled to drive their children to school and other places because of their uncertainty about what other parents are deciding to let their children do. The situation reduces the freedom of both parent and child to use the healthiest and safest means of travel for any given circumstance. Tranter outlines strategies that may allow parents to escape social traps and hence allow their children more freedom to explore their own neighbourhood and city. Finally, Prue Walsh (Chapter 9) draws on her practical experience in recreation space design to assess the efforts of planners, developers and designers to provide recreational spaces for children in cities. She finds all three professions wanting, pointing to the many failures in planning, design and development of this most critical urban sphere of action for children. Walsh offers two practical sources of advice for practice improvement: a set of planning guidelines and a design brief that, taken together, provide a sophisticated set of pointers for the creation of good urban playspaces.

The final chapter in the volume, 'Pathways to the child friendly city', draws on the propositions of preceding chapters. This chapter aims to synthesize the material from each of the contributing authors into a framework for creating more child friendly cities. The aim here is not to outline the entire range of actions and improvements needed to create child friendly cities, but to distil the particular wisdom offered by the specialist contributors to this volume. Creating child friendly cities is a vast enterprise that will involve many forms of expertise, and many lines of intervention which cannot be addressed in the pages that follow. Our hope is that this book makes a genuine and original contribution to the urgent project that confronts us: creating child friendly cities.

Note

1 Dr Fiona Stanley AC was Australian of the Year in 2003.

References

Cadzow, J. (2004) 'The bubble-wrap generation', *The Sydney Morning Herald*, 17 January: 18–21.

Christensen, P. and O'Brien, M. (eds) (2002) *Children in the City: Home Neighbourhood and Community (The Future of Childhood)*, London: Routledge.

Eberstadt, M. (2004) *Home-Alone America: The Hidden Toll of Day Care, Behavioral Drugs, and Other Parent Substitutes*, London: Sentinel HC.

Gump, P. V. (1975) *Ecological Psychology and Children*, Chicago, IL: University of Chicago Press.

Homel, R. and Burns, A. (1985) 'Through a child's eyes: quality of neighbourhood and quality of life', in Burnley, I. and Forrest, J. *Living in Cities: Urbanism and Society in Metropolitan Australia*, Sydney: Allen & Unwin.

Lady Allen of Hurtwood (1968) *Planning for Play*, Norwich: Jarrolds Publishing.

Luthar, S. (2003) 'The culture of affluence: psychological costs of material wealth', *Child Development*, 74 (6): 1581–93.

Lynch, K. (1977) *Growing Up in Cities*, Cambridge, MA: MIT Press.

Stanley, F., Richardson, S. and Prior, M. (2005) *Children of the Lucky Country? How Australian Society Turned its Back on Children and Why Children Matter*, Sydney: Pan Macmillan.

Ward, C. (1978) *The Child in the City*, London: The Architectural Press.

Part One

Scales of analysis

Chapter 2

United Nations

A key player in a global movement for child friendly cities

Karen Malone

Introduction: a global cry for action

> Children are not only our future, they are our present and we need to start taking their voices very seriously.
>
> Executive Director of UNICEF (United Nations Children's Fund) Carol Bellamy (UNICEF 2001a)[1]

Worldwide, many things have impacted on children's health and well-being. Industrialization, population growth, poverty, environmental degradation, crime and war, and the constant dumping of toxic waste into the atmosphere, the waterways and the soil have all played their part. During the past 50 years population growth and resource consumption have both been increasing exponentially. Consequently, cities have been growing at alarming rates. Almost two billion people worldwide live in urban regions of the developing world. This figure is projected to double over the next 30 years, at which time urban dwellers will account for nearly half the global population (UN 2003). Moreover, most of these urban dwellers are likely to be living in slums, which will result in the 'urbanization of poverty'. According to the United Nations (2003) report on its progress towards its Millennium Development Goals, *Improving the Lives of 100 Million Slum Dwellers*: 'Slums are a physical and spatial manifestation of increasing urban poverty and intra-city inequality.' UN (2003) estimates based on figures from the Global Urban Observatory demonstrate that 31.6 per cent of the world's urban population, or 924 million people, are living in slums. Breaking this down further to least developed countries (LDCs), the number increases dramatically to 78.2 per cent. This means that 140 million people of 179 million urban dwellers are living in slums. With the world's largest cities (megacities) growing by over one million people per week (Satterthwaite 1996; United Nations Centre for Human Settlements (UNCHS) 1996), it has been estimated that by the year 2025 the world's largest cities would need to accommodate four billion people (UNCHS 1996). With an average of one-third of the population in developed nations consisting of children under the age of 18 years, the majority of these city residents (at least 45–50 per cent) will be children. This number increases in low-income nations by up to 60 per cent (Dallape 1996):

> From the point of view of poverty defined by monetary income, children are over-represented in relation to both the total number of poor people and their age groups. Estimates carried out using the international poverty line show that around half of the poor population is made up of children.
>
> (Minujin, Vandemoortele and Delamonica 2002: 31–2)

Rapid urban growth creates huge imbalances between available resources and the needs of the population. With one-third of children in the developing world already living in sub-standard housing or homeless (UNICEF 2001a), this situation is bound to get worse as cities become larger and resources scarcer. As Kofi Annan, Secretary-General of the UN, clearly articulates in the foreword of the *State of the World's Children* document for 2005:

> The State of World's Children 2005 makes clear, for nearly half the two billion children in the real world, childhood is starkly and brutally different from the ideal we all aspire to. Poverty denies children their dignity, endangers their lives and limits their potential. Conflict and violence rob them of a secure family life, betray their trust and their hope. HIV/AIDS kills their parents, their teachers, their doctors and nurses. It also kills them. With childhood of so many under threat, our collective future is compromised. Only as we move closer to realizing the rights of all children will countries move closer to their goals of development and peace.
>
> (UNICEF 2005)

The consequences on children's lives of not planning for the future are clear. If cities do not address ways of growing in sustainable ways, and provide adequate infrastructure to support population growth, the impact will be the continuance or exacerbation of large-scale poverty and urban slums. It is for these reasons that action based on the principles of sustainable development and children's rights has been launched by the UN.

UN focus on sustainable development and children's rights

The principles of sustainable development clearly demand that the achievement of environmental, social and economic goals meet the needs of the present generation without compromising future generations. Nation governments must maintain the integrity of the social, economic and environment fabric of their global and local environments through processes that are participatory and equitable. The principles of the CRC (UNICEF 1992) reinforce this responsibility of the states. The convention challenges them to uphold the child's right to live in a safe, clean and healthy environment, and to engage in free play, leisure and recreation in this environment. According to the CRC, a child's well-being and quality of life is the ultimate indicator of a healthy environment, good governance and sustainable development (UNICEF 1992; UNICEF 1997). If these goals of sustainability are not achieved, children will be more profoundly affected than other members of the global community. Clearly, a convergence of the UN principles of sustainable development and of children's rights has provided the foundation for the child friendly city global movement.

This connection between children's rights and sustainable development has been formally articulated in a number of UN global declarations and documents emerging from intergovernmental summits and meetings. Some of the most significant documents for stimulating discussions on children and sustainable development include the *Plan of Action* (UNICEF 1990). This resulted from the World Summit for Children, the Rio Declaration and the action plan of *Agenda 21,* endorsed at the Earth Summit in Rio de Janeiro in 1992 (UN 1992). Principle 21 of the Rio declaration clearly reinforces the active participatory role of youth in sustainable development: 'The creativity, ideals and courage of the youth of the world should be mobilized to forge a global partnership in order to achieve sustainable development and ensure a better future for all' (UN 1992). The introduction and the content of Chapter 25.1 state: '[y]outh comprise nearly 30 per cent of the world's population. The involvement of today's youth in environmental and developmental decision-making, and in the implementation of programmes is critical to the long term success of Agenda 21' (UN 1992).

More recently, the Rio+10 conference in Johannesburg in 2002 renewed its commitment to sustainable development and the goal of young people's participation as a key to making the vision a reality. The commitment by the UN to launch the Decade of Education for Sustainable Development in 2005 presents an ongoing focus on the active participation of children, youth and all community members in developing social, cultural and educational strategies to address sustainability issues.

The projected impact of urbanization on human quality of life was the focus of the first Habitat Agenda proclaimed in 1976. The agenda was revisited 20 years later at the United Nations Conference on Human Settlements at Istanbul in 1996. From this meeting emerged the *Habitat II Agenda* (UNCHS 1997). A delegation from UNICEF presented the draft 'Children's Rights and Habitat' at the meeting. This document drew specific attention to the important role children have in sustainable development, and launched the term 'child friendly cities'. From UNICEF's report, and after much lobbying from conference participants, a preamble was inserted into the *Habitat II Agenda* that included a specific focus on children and youth:

> The needs of children and youth, particularly with regards to their living environment have to be taken fully into account. Special attention needs to be paid to the

> participatory processes dealing with the shaping of cities, towns and neighbourhoods; this is in order to secure the living conditions of children and of youth and to make use of their insight, creativity and thoughts on the environment.
>
> (UNCHS 1996)

After the Istanbul conference in 2001, a reconfirmation of the principles of the *Habitat Agenda* was again taken up at the Habitat +5 meeting in New York. At the local level, the goals of sustainable development and children's rights are also expressed through *Agenda 21* (UN 1992). This is the action plan for local governments, communities and all stakeholders to promote and implement sustainable development. UNICEF has utilized *Agenda 21* to support the Child Friendly Cities Initiative (CFCI) as a programme of action, encouraging mayors and community organizations to involve children in partnerships on environmental decision-making.

As this groundswell of UN policy continued to recognize the impact of urbanization on children's lives, and the importance of immediate actions in addressing this impact, a plan was put forward to revive the UNESCO Growing Up In Cities (GUIC) project. GUIC was officially launched in 1996. Its goal was to utilize the UN policy framework on sustainable development and children's rights. The aim was to design a global model of participatory research and action, where children and young people evaluated the quality of city environments and engaged in collaborative projects with local officials to initiate actions for change. Additionally, a team of multidisciplinary social and physical researchers sought to develop a set of qualitative indicators of quality of life. Designed with children involved, these indicators were seen as having the potential for supporting local government officials in their attempts to construct baseline data on how child friendly their cities were (see Table 2.1).

The inclusion of these specific documents and the policy framework produced, was an attempt to ensure policy makers devoted special attention to the needs of children, sustainable environmental and citizens engaging in active participation. Children are acknowledged as having both the greatest stake in long-term environmental stability and the capacity to act as protagonists in achieving that stability (Bartlett *et al.* 1999). These global initiatives provided a framework for supporting policy development. The test for local governments was to decide how to put them into action. Local governments have been charged with the task of ensuring that the principles of *Agenda 21* and the spirit of the CRC (UNICEF 1992) are the impetus to create appropriate mechanisms for children's participation in building a sustainable and equitable urban future.

The relationship between sustainable development and children's lives is not just about adults' roles as stewards and their capacity to act on behalf of

Table 2.1 Indicators of local environmental quality (based on the evaluations of 10–15-year-olds at 'Growing Up In Cities' sites)

	Social qualities	*Physical qualities*
Positive	Social integration	Green areas
	Freedom from social threats	Provision of basic services
	Cohesive community identity	Variety of activity settings
	Secure tenure	Freedom from physical dangers
	Tradition of community self-help	Freedom of movement
		Peer gathering places
Negative	Sense of political powerlessness	Lack of gathering places
	Insecure tenure	Lack of activity settings
	Racial tensions	Lack of basic services
	Fear of harassment and crime	Heavy traffic
	Boredom	Trash/litter
	Social exclusion and stigma	Geographic isolation

Source: taken from Chawla 2002.

the child, it is also about recognizing the capacity for children and youth to be *authentic participants* in planning, development and implementation processes (UNICEF 1997). Two key UN global child-centred projects, CFCI and GUIC, were strongly influenced by the UN policy climate of the day. This led to their simultaneous launch in 1996. CFCI focused on city officials, and the development of a plan of action to evaluate and develop ongoing strategies to transform cities into child friendly environments. GUIC provided the tools for the governments and their communities to engage children actively in the evaluation process. The next sections will look at these two projects in more detail.

UNICEF's child friendly cities

> A child friendly city is a system of good local governance committed to the fullest implementation of the Convention on the Rights of the Child. Large cities, medium-size towns as well as smaller communities – even in rural settings – are all called to ensure that their governance gives priority to children and involves them in decision-making processes.
>
> (UNICEF 2004a: 1)

UNICEF's CFCI as defined above has been building a global network of resources in cities for the past ten years on almost every continent on the earth. But how did it come to be? This section starts with a brief history of CFCI and then describes in more detail the programme characteristics.

In Istanbul on 5 June 1996, during the International United Nations Habitat II conference, a delegation of international practitioners, researchers and child activists delivered a document. This was in response to an overwhelming recognition during preparatory meetings that insufficient attention had been given to the issues of safe, secure and healthy living conditions for children. The document was entitled 'Children's Rights and Habitat'. At a workshop organized by UNICEF during this conference, participants used the framework of the document to build a detailed report (of the same name) that identified children's rights and the conditions for achieving them in a highly urbanizing world. They also sought to identify local and regional obstacles to achieving what was defined as the child friendly city. Additionally, after overwhelming response, it was acknowledged that bringing children's needs and concerns to attention meant creating a world that was better for all inhabitants. That is, a child friendly city is a people friendly city (UNICEF 1996). During the conference and the lead-up to the major event, it was declared that: 'the well-being of children is the ultimate indicator of a healthy habitat, a democratic society and good governance' (UNICEF 1996). What the delegation was able to do at the conference was to remind participants of the CRC, and the importance of evaluating past, present and future cities according to these statutory rights. UNICEF's CFCI was first conceived in response to the United Nations Conference on Environment and Development in 1992. It came at a time when the situation of urban children around the world was recognized to be of critical concern. Discussions on sustainable development, the management of human settlements and the rights of children could not be done in isolation. The emerging child friendly cities philosophy was underpinned by the view that to actively implement the CRC at a national and local government level a healthy environment, good governance and sustainable development are necessary (UNICEF 1992; UNICEF 1997).

The *Children's Rights and Habitat Report* (UNICEF 1997) became the first step in the emerging CFCI programme. It was presented by delegates of UNICEF to participants at the United Nations Conference on Human Settlements at Istanbul in 1996. Its main role was to draw attention to the need to support children. 'The needs of the children and youth, particularly with regard to their living environment, have to be taken fully into account' (UNICEF 1997: 13), and to acknowledge the role children have in being active participants in creating these sustainable human settlements. 'Children have a special interest in the creation of sustainable human settlements that will support long and fulfilling lives for themselves and future generations. They require opportunities to participate and contribute to a sustainable urban future' (UNICEF 1997: preamble).

The CFCI has emerged in recognition of several worldwide trends. These include 'the rapid transformation and urbanization of global societies; the growing responsibilities of municipal governments and community for their populations in the context of decentralization; and consequently, the increasing importance of cities and towns within the national political and economic systems' (UNICEF 2004a: 1). The guiding principle behind the initiative is that safe and supportive environments nurture children of all ages with opportunities for recreation, learning, social interaction, psychological development and cultural expression, and promote the highest quality of life for its young citizens.

Building on the work developing from the launch of the CFCI in the late 1990s, the CFCI was recognized as being a critical programme for supporting the outcome document of the UN General Assembly's Special Session on Children in May 2002, *A World Fit for Children* (UNICEF 2002a). This document identified the importance of local government and authorities in creating partnerships to promote and protect the rights of children, and the significance of building on the CFCI . The document states:

> local governments and authorities can ensure that children are at the centre of agendas for development. By building on ongoing initiatives such as child friendly communities and cities without slums, mayors and local leaders can significantly improve the lives of children.
>
> (UNICEF 2002a)

A media release on 7 February 2003 from the UNICEF Innocenti Research Centre (the home of the International Secretariat for the Child Friendly Cities Initiative) in Florence confirmed the significant role of CFCI in addressing children's specific needs:

> The tens of millions of urban children who are denied basic social services – such as education and health care – are living proof that the world has systematically failed to protect them. These children deserve to live in a protective environment – one that safeguards them from abuse and exploitation. This was the commitment reaffirmed by the Heads of State and Government in 2002, at the Special Session on Children and we need to take it seriously and translate it into action.
>
> (UNICEF 2002b)

Child friendly cities also featured widely in other documents emerging from UNICEF around this time, including the *Partnerships to Create Child Friendly Cities* (2001a) and *Poverty and Exclusion among Urban Children* (2002b).

The CFCI has been enhanced through its partnership with the International Union of Local Authorities (IULA) who, with UNICEF, shares a common interest in the importance of supporting children and women in local contexts throughout the world. IULA is a key stakeholder in the CFCI programme, as many local governments carry the final responsibility for the very elements that have the greatest impact on children's well-being and quality of life – education, health, housing, environmental protection, recreation and transport. This builds on their previous efforts, which go as far back as 1992, when the Mayors, Defenders of Children Initiative was launched. Mayors and other leaders pledged to make children's basic needs a priority, to increase children's participation, to review and revise legislation, and to assist children affected by war and other adverse circumstances (UNICEF 1997). The IULA has strongly encouraged its members, 112 national associations of local authorities and 200 individual cities in more than 110 countries, to take up the role of ambassadors for the CFCI.

So what are the characteristics of a child friendly city? How do you become recognized as a UNICEF child friendly city? A conceptual paper was written for the International Child Friendly Cities Secretariat by Peter Newell in 2003, in preparation for the European cities workshop held at the Innocenti Research Centre. The paper states that the process of building a child friendly city:

> is the process of implementing the Convention on the Rights of the Child led by local government in an urban context. The aim is to improve the lives of children now by recognizing and realizing their rights – and hence transform for the better urban societies today and for the future. Building child friendly cities is a practical, not theoretical, process which must engage actively with children and their real lives.
> (Newell 2003: 2)

According to UNICEF (2001a) a child friendly city requires basic elements that ensure it is able to fulfil the principles of the CRC. These are listed in Table 2.2.

According to the CFCI website (UNICEF 2004b), 'A Child Friendly City is a local system of good governance committed to fulfilling children's rights.' It does this by being actively engaged in fulfilling the right of every young person to:

- influence decisions about their city;
- express their opinion on the city they want;
- participate in family, community and social life;
- receive basic services such as health care and education;
- drink safe water and have access to proper sanitation;
- be protected from exploitation, violence and abuse;
- walk safely in the streets on their own;
- meet friends and play;
- have green spaces for plants and animals;
- live in an unpolluted environment;
- participate in cultural and social events;

Table 2.2 Characteristics of a child friendly city

- Good access for all children to affordable, quality basic health services, clean water, adequate sanitation and solid waste removal;
- Local authorities to ensure that policies, resources allocations and governance actions are made in a manner that is in the best interests of the children and their constituencies;
- Safe environments and conditions that nurture the development of children of all ages with opportunities for recreation, learning, social interaction, psychological development and cultural expression;
- A sustainable future under equitable social and economic conditions, and protection from the effects of environmental hazards and natural disasters;
- Children have the right to participate in making decisions that affect their lives and are offered opportunities to express their opinions;
- Special attention is given to disadvantaged children, such as those who are living or working on the streets, sexually exploited, living with disabilities or without adequate family support;
- Non-discrimination based on gender, ethnic background or social or economic status.

Source: UNICEF 2001a: 3.

- be an equal citizen of their city with access to every service, regardless of ethnic origin, religion, income, gender or disability.

There is no single definition of what a child friendly city is or ought to be. In fact the documents go to great length to say that cities can never achieve child friendly status because they will always be transforming and responding to the changing local and global context. In some cities, especially in high-income nations, emphasis tends to be on environmental and physical issues such as improving recreational spaces and green spaces, alienation and controlling traffic to make streets safe for young citizens. In low-income nations, the focus is frequently on increasing access to basic services for the poor.

UNICEF's CFCI Secretariat has developed a toolkit to support cities in working towards achieving child friendly cities. This toolkit is available electronically via its website, and it has six key elements or resource banks, including: CFCI framework; lessons from CFCIs; partnerships and networking; the nine building blocks; the CFCI database; and the CFCI Secretariat (see Figure 2.1).

The key components of this toolkit are nine building blocks. They act as guidelines to be used by government, and are essential if they are to engage children in authentic participatory processes. They also ensure children's rights are incorporated into all levels of relevant decision-making forums, particularly in terms of the equitable distribution of basic services. The framework outlines

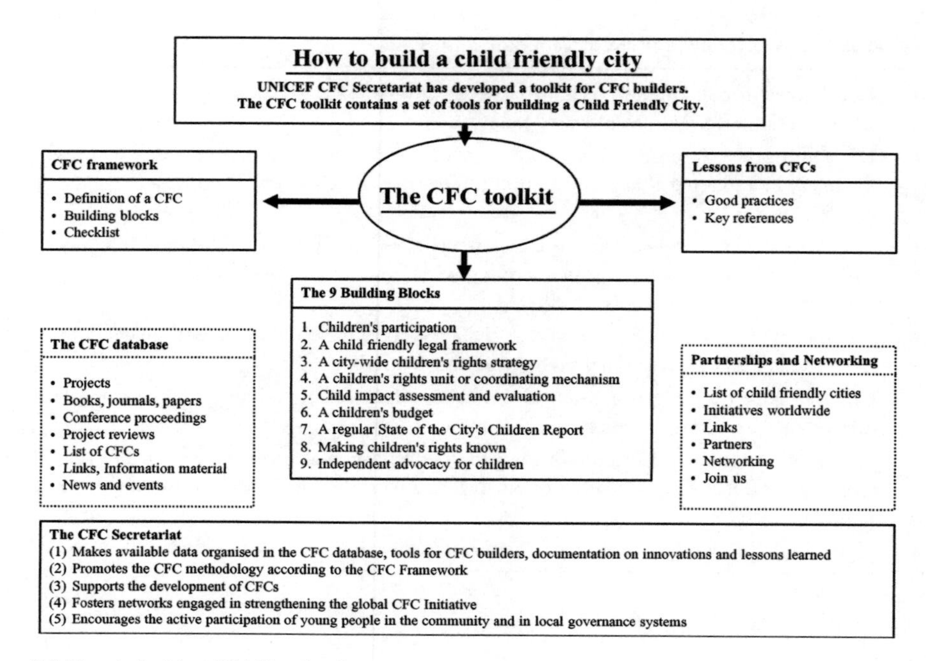

2.1 How to build a child friendly city.

Source: UNICEF (undated).

the fact that to make change happen, governments must make political commitments and implement actions on the ground. Table 2.3 provides an overview of the nine elements or building blocks for developing a child friendly city.

The document written to support government agencies and/or associate bodies when implementing these nine elements – 'Building blocks for developing a child friendly city' in the publication *Building Child Friendly Cities: A Framework for Action* (UNICEF 2004a) – includes a definition for each of the elements and a checklist for evaluating activities and consistency in achieving them.

The International Secretariat for Child Friendly Cities strives to act as a resource and facilitator for the toolkit activities. The experience of the Secretariat, according to Riggio (2002: 58), is that the success of the development of the initiatives are dependent on 'the capacity of cities to make links with each other and to share resources, in order to multiply the effects of efforts that are still limited in relation to need'. She goes on to state:

> Operating at the crossroads of information sharing perhaps the Secretariat's most critical role is as a common reference point for child friendly cities, not to guide programmes but to support their growth and facilitate the exchange of experience and knowledge.
>
> (2002: 58)

Table 2.3 The nine building blocks for developing a child friendly city

1	Children's participation
2	A child friendly legal framework
3	A city-wide children's rights strategy
4	A children's rights unit or coordinating mechanism
5	Child impact assessment and evaluation
6	A children's budget
7	A regular State of the City's Children Report
8	Making children's rights known
9	Independent advocacy for children

Source: UNICEF 2004a: 4.

In this capacity the International Secretariat of CFCI has been undertaking a review of CFCI strategies from the around the world, adopting a common research protocol on the basis of the CFCI framework for action. The key focus areas used as major headings for review include situation analysis, child and youth participation, cross-sectoral approaches, strategic partnerships, attention for disadvantaged groups, linkages with CFCI networks, advocacy strategies, resource mobilization, institutional and legal reforms, capacity building, and monitoring and evaluation. The case studies contain distilled information on methods and approaches used by CFCI projects and a set of good practices and lessons learnt. The case studies are across high-, middle- and low-income nations. One of the key projects evaluated by the CFCI team and identified as a model for supporting cities to begin their CFCI process has been the UNESCO GUIC project.

UNESCO Growing Up In Cities project

If having rich environmental experiences and feeling safe and secure, connected and valued are universal indicators of quality of life, then what better place to start than to evaluate cities through the eyes of its children? The UNESCO GUIC project has been doing this for the past ten years. GUIC involves children worldwide in the process of evaluating their urban environments and improving the conditions of their lives. Figure 2.2 is a drawing by four-year-old Sara who, as a member of a children's UNESCO research team, has designed some alternative uses for an old abandoned convent site near her high-rise commission flats. Sara's drawing, submitted to the Melbourne State Government as part of a local community submission to stop the overdevelopment of a local parkland and heritage site, illustrates that there is no age limit on creativity and authentic participation in the evaluating and planning of the physical environment. This is the ultimate goal of the GUIC project – influence municipal policies through the inclusion of children's perspectives and build alliances and shared actions between people who are supporting children's rights in community-based and non-government agencies across a variety of local, national and international government bodies. Coordinated from UNESCO headquarters in Paris the project started as a revisitation of an earlier project by Kevin Lynch in the early 1970s. The project has expanded extensively from its original eight sites in 1996 to over 30 sites in 2005. There have been two key outcomes from the original revisit of eight countries, using and building on Lynch's original research tools. The first is the publication of a book about the project and stories from the field (Chawla 2002). The second is a participatory manual, *Creating Better Cities with Children and Youth* (Driskell 2002), which provides a toolkit of principles, research methods and strategies for working in authentic participatory ways with children. The manual has become the catalyst for a number of workshops for municipal councils, urban planners, teachers and child researchers that have been staged around the world. GUIC, through its participatory work with children and youth, also identified a list of child-generated and culturally pluralistic indicators of quality of life (Chawla 2002). These indicators have now been used to support the compilation of city-wide initiatives to define and support the child friendly cities initiatives (see Table 2.1).

The list of positive indicators for urban environments identified by children in cities throughout the world included such elements as provision for basic needs, social integration, safety and free movement, peer gathering places and safe green spaces. The negative indicators include social exclusion, violence and crime, heavy traffic, lack of gathering places, boredom and political powerlessness (Chawla 2002). Having a child friendly city, according to the work conducted by the GUIC teams, means that the social and physical environment allows children to feel a sense of belonging, to be respected and valued, and to have opportunities to become increasingly independent (Hart 1997; Malone and Hasluck 2002). Children also voiced the importance of having easy access to safe areas for socializing and playing with their friends, as well as a continued desire for green or 'wild' spaces for contact with nature (Chawla 2002; Malone and Tranter 2003). What was also clear from the global comparisons of low- and high-income nations was the difference in children's specific 'concerns'. Children in low-income nations (such as India, South Africa, Argentina, Papua New Guinea), who were clearly impoverished and in need of basic infrastructure, more often spoke of their rich, diverse and engaging social environment than did children from high-income nations. That is, for many children a positive city environment was one where they felt a sense of belonging, pride, responsibility and of being cared for by their families and the community. Children in high-income nations by comparison (US, UK and Australia) were also inclined to focus on the social aspects of the environment, but spoke of their own impoverishment due to the marginalized position they often found themselves in. Feelings of alienation, lack of family and community support, and a general sense of despondency about their future meant

2.2 An example of the input of a young child (aged four years) to the redesigning of a shared community space in Collingwood, Melbourne, through a drawing.

Source: UNESCO GUIC Asia-Pacific and Karen Malone.

that many young people were quite negative about their urban environments. Of course, this is an oversimplification. For example, for the children in the camps of Johannesburg, the detrimental impacts of a lack of secure tenure and no basic services mean they carry the physical and social stigma of being 'squatters' open to discrimination and harassment (Swart-Kruger 2002).

Clearly, one of the substantial outcomes of the ongoing GUIC project has been the recognition and practical application that young people need to be involved in determining how their communities work for them. Young people's views on community problems and resources often differ from the ones adults construct for them. GUIC provides a framework for supporting policy development – the test for local governments is to put them into action. Responding to children's rights is not just about adults' roles as stewards. It is also about recognizing the capacity for children and youth to be *authentic participants* and the importance this has on their continued sense of connection to their community (Malone 1999). Democratic behaviour is learnt through experience; it is imperative that children be given a voice in their communities so that they will be able to participate fully in civil society (UNICEF 1997; Malone and Hasluck 1998). It was clear from my own GUIC research work (I have been the Asia-Pacific Director of GUIC since its onset in 1996) that much that passes as children's participation in government processes is nothing more than mere tokenism, particularly when many city officials are convinced they already know what children need. The commitment of the GUIC project teams has been to genuine participation, with increasing degrees of self-directed and child-initiated participation (see Hart 1997). Supporting children to help them take on responsibility and ownership of projects has a two-fold advantage. First, it develops the self-esteem and confidence of individuals to take up leadership roles. Second, it increases project sustainability. It also draws on natural curiosity, creativity and children's skills (particularly in media and technology) – so the project can evolve in ways never imagined by the adult facilitators. Children researching through video interviews with their peers and other community members (as illustrated in Figure 2.3) and turning these interviews into short films is one example of self-directed participation. Other examples are documented in the GUIC South African (Swart-Kruger 2002) and Argentinian sites (Cosco and Moore 2002).

Finding ways to work authentically with children, and allow their voices to be heard is clearly one of the key outcomes of the GUIC project in terms of contributing to the CFCI programme and actualizing CRC at a local level. Because, as Chawla (2002: 234) notes in the final chapter of the GUIC book, participation is not just a social but a very personal act, '[p]articipation may serve extrinsic purposes in terms of concrete community improvements and more child-sensitive policies, but it also appears to foster an intrinsic sense of self-esteem and self-efficacy that is a basic preparation for citizenship'.

2.3 An example of young people's authentic participation: these youths are interviewing their peers using videos in order to present to the local council issues surrounding the quality of their neighbourhood, Braybrook, Melbourne.

Source: UNESCO GUIC Asia-Pacific and Karen Malone.

Conclusion: global commitments to child friendly cities

Cities, as a response to the changing global context, can provide opportunities to support children if they are managed using the principles of good governance advanced through the UN policy frameworks. These are equity, social inclusion, accountability and a commitment to children's rights. In the absence of this 'good governance' and a lack of investment in infrastructure, social services, waste and resource management, urban settlements can become life-threatening environments for children and their families. But in terms of addressing the complex needs of an urbanizing world, cities are a key to providing the physical and social needs of the escalating child population. And if cities are built with children's needs in mind, there are a number of potentially positive advantages for urban children compared to their rural counterparts. Due to increased density, the number, variety and quality of public and private services can be greater in terms of cultural, commercial, recreational, health, educational, psychological support, religious, and municipal services (Churchman 1999). The variety of these services may also be greater and therefore cater to the diversity of children's needs. Cities also provide significant economies of scale and proximity for very basic services such as water, electricity, sewage and communication. Distances within the neighbourhood, or within other parts of the city, may be less and therefore enhance the possibility of children's mobility and independence, and facilitate opportunities to reach a range of resources by walking or bicycle riding. Public transportation is also likely to be more available, and potentially more accessible, comfortable, frequent and affordable, in a city landscape. The people whom children come in contact with also offer diverse opportunities for further support and for building community engagement. Furthermore, the variety of stimuli – sensory, cognitive, social, environmental and emotional – is likely to be greater in cities than in other kinds of settlements (Gehl 1996). All of these could be considered to be positive aspects of the city. They have, through the work of both the CFCI and GUIC project, been shown to represent advantages and opportunities for children.

On the other hand, many of these advantages, if not managed well, can become disadvantages or limit children's lives. Issues such as the danger of violence from adults or other children and exposure to other criminal acts can put children at risk when moving around the city (Carbonara-Moscatti 1985; Spencer and Woolley 2000; Malone and Hasluck 2002). High levels of motorized traffic can endanger children from road accidents, as can the risk of noise and air pollution (Gehl 1996; Stansfield, Brown and Haines 2000).

Children are vulnerable in city environments where family breakdown, poverty, neighbourhood conflicts, vandalism and bullying in schools have restricted their access to local social and physical resources. One consequence of the impact of urban risk on children's lives is their retreat to 'home environments', with many parents often prescribing and circumscribing children's access to the environment (Malone and Hasluck 2002). Spontaneous unregulated play in neighbourhood spaces, particularly in affluent areas of cities, is increasingly becoming a thing of the past. Children are encouraged to participate in regulated play environments in their homes, friends' houses and commercial facilities (Malone 1999; McKendrick, Bradford and Fielder 2000; Malone and Hasluck 2002). This type of regulatory practice may help to 'protect' children from becoming victims of environmental hazards, but it also has long-term consequences for their social and emotional growth. The view of the 'dangerous city' fed by the current climate of fear sweeping the Western world adds to the circumventing of children's experience and independence in their urban environment. It has also contributed to a lack of opportunities for children to participate in city life. Research across all the GUIC sites has repeatedly illustrated that discovering how to negotiate the social and environmental context of a neighbourhood is important for developing children's independence, resilience and social competence. But a safe, child friendly environment is needed before these important opportunities are going to be freely available to all children: 'Safe environments nurture children of all ages with opportunities for recreation, learning, social interaction, psychosocial development and cultural expression' (UNICEF 2001b).

Recent research emanating from the CFCI and GUIC projects (Satterthwaite *et al.* 1996; UNICEF 1997; Malone 2001; Chawla 2002; Malone and Hasluck 2002; Swart-Kruger 2002) reveals that children and youth around the world have the same needs no matter where they grow up. They articulate to adults a clear message of what they need in order to be able to function adequately in a city environment. They want:

- clean water and enough food to eat;
- to be healthy and have the space to learn, develop and play;
- friends and family who love and care for them;
- to participate in community life and be valued;
- to collaborate with adults to make the world a better place for all;
- peace and safety from threats of violence;
- access to a clean environment where they can connect with nature; and
- to be listened to and their views taken seriously.

These positive and negative characteristics of cities become even more important as children grow older. During the earlier years, they are more dependent upon adults and less tied to the limits of their neighbourhood environment. However, at an older age, they seek to expand their use of the environment beyond the neighbourhood alone and to reap more of the benefits the city has to offer.

There are many probable implications for the lives of children, if they are not able to function relatively independently as they grow up. These could include:

1 Limitations to their capacity to expand their environmental literacy. We know from research that independent experience in an environment increases environmental literacy, capabilities and confidence. The more complex the environment becomes, the more capabilities are required, and the more it becomes an essential life skill.
2 A lack of experience of autonomy and independence that means they are not allowed to see that they are capable of functioning on their own, thus contributing to their feelings of self-confidence and self-esteem (Lang and Deitz 1990).
3 An inability to obtain different kinds of privacies or freedoms of choice to reinforce their feelings of self-worth and efficacy.
4 A lack of opportunity, due to restricted mobility, to make use of the diversity of people, environments, activities, resources and stimuli that the city offers.
5 Limitations in their exposure to measured risks and challenges that would allow them to learn how to become streetwise and knowledgeable about how to function as an active and sociable participant in their city environment (Hart 1997).

When children cannot function independently and are unable to take advantage of what the city has to offer them, they will have limited experiences and less opportunity to enhance their own lives or make a contribution to the lives of others. Whether recognized as an economic, social or cultural benefit, having cities where children are independent, mobilized and contributing to local capacity is a long-term investment in the life of that community.

Cities, as found in the work of the GUIC team globally (Chawla 2002), are often designed as places where adults go about their business. They drive, walk, work, park their cars, shop and visit restaurants to meet friends. Children's places within the city are relatively confined and essentially invisible, and this trend is accelerating if the comparisons from the 1972 evaluation of neighbourhoods in Melbourne are treated as broad general evidence (Malone and Hasluck 2002).

Many of the child friendly cities described in the case studies of the UNICEF CFCI network certainly paint a very different picture of childhood and the way children are positioned in the city landscapes. What seems to make these cities unique is the way they view children as an investment in the shared future of their community, and the extent to which they support the key principles of sustainable development and children's rights.

Children are the canaries in the mines. Youth suicides, drug use, and feelings of disenfranchization and disconnection are increasing in epidemic proportions in cities across Australia and in many nations around the world. As the world becomes more urbanized, and the pressures and demands on resources become greater, there will be a stronger need to develop strategies and plans that have their basis in global policies and frameworks that countries are signatories to, such as *Agenda 21* and the CRC. These UN declarations become political devices for making the State accountable for the ways they plan and design our city landscapes. Unless this becomes part of mainstream government practice, our young people will continue to be susceptible to the escalating disadvantages of the cities.

In summary, cities can be positive and negative places for children, depending on whether they are the wealthy few or the impoverished many. Ideally, neighbourhoods, towns and cities should be places where children can socialize, observe and learn about how society functions and contribute to the cultural fabric of their community. They should also be sites where they find refuge, discover nature and find tolerant and caring adults who support them. For cities to be supportive of children's needs and to fulfil their obligations in terms of the principles contained in the CRC (UNICEF 1992), sustainable systems and increased local capacity are required. These will only be possible by creating environments that are based on cooperation and partnerships at a variety of different levels across national and local borders, social domains and groups. UNICEF's CFCI, and its companion project, UNESCO GUIC project, provide a global framework for mobilizing policy into action. If only adults were as optimistic as children:

> the children's view of the city is optimistic and full of life, open to the future while firmly rooted in the present. A sense of the future that boldly demands to be listened to and dialogued with. A tenacious feeling of optimism that claims the right to be part of the dialogue in giving shape and identity to the city.
>
> (Davoli and Fari 2000: 18)

Note

1 United Nations International Children's Emergency Fund (UNICEF): the UNICEF website describes it as the driving force that helps build a world where the rights of every child are realized (www.unicef.org/about/who/index.html, accessed 4 May 2006).

References

Bartlett, S., Hart, R., Satterthwaite, D., De La Barra, X. and Missair, A. (1999) *Cities for Children: Children's Rights, Poverty and Urban Management*, London: UNICEF/Earthscan.

Carbonara-Moscatti, V. (1985) 'Barriers to play activities in the city environment: a study of children's perceptions', in Garling, T. and Valsiner, J. (eds) *Children Within Environments*, New York: Plenum.

Chawla, L. (2002) *Growing Up in an Urbanizing World*, London: UNESCO/Earthscan.

Churchman, A. (1999) 'Disentangling the concept of density', *Journal of Planning Literature* 13 (4): 389–411.

Cosco, N. and Moore, R. (2002) 'Our neighbourhood is like that!', in Chawla, L. (ed.) *Growing Up in an Urbanizing World*, London: UNESCO/Earthscan.

Dallape, F. (1996) 'Urban children: a challenge and an opportunity', *Childhood*, 3: 283–94.

Davoli, M. and Fari, G. (2000) *Reggio Tutta: A Guide to the City by Children*, Reggio Emilia, Italy: Reggio Children.

Driskell, D. (2002) *Creating Better Cities with Children and Youth: A Participatory Manual*, London: UNESCO/Earthscan.

Gehl, J. (1996) *Life Between Buildings*, Copenhagen: Arkitektens Forlag.

Hart, R. (1997) *Children's Participation*, London: Earthscan/UNICEF.

Lang, M. and Deitz, S. (1990) 'Creating environments that facilitate independence: the hidden dependency trap', *Children's Environments Quarterly*, 7 (3): 2–6.

Malone, K. (1999) 'Growing Up In Cities as a model of participatory planning and "place-making" with young people', *Youth Studies Australia*, 18 (2): 17–23.

Malone, K. (2001) 'Children, youth and sustainable cities' (editorial special issue), *Local Environment*, 6 (1): 5–12.

Malone, K. (2002) 'Street life: youth, culture and competing uses of public space', *Environment and Urbanization*, 14 (2): 157–68.

Malone, K. and Hasluck, L. (1998) 'Geographies of exclusion: young people's perceptions and use of public space', *Family Matters*, 49, Autumn: 20–6.

Malone, K. and Hasluck, L. (2002) 'Australian youth: aliens in a suburban environment', in Chawla, L. (ed.) *Growing Up In an Urbanizing World*, London: UNESCO/Earthscan.

Malone, K. and Tranter, P. (2003) 'School grounds as sites for environmental learning: making the most of environmental opportunities', *Environmental Education Research*, 9 (3): 283–303.

McKendrick, J., Bradford, M. and Fielder, A. (2000) 'Kid customer? Commercialisation of playspace and the commodification of childhood', *Childhood*, 7 (3): 295–314.

Minujin, A., Vandemoortele, J. and Delamonica, E. (2002) 'Economic growth, poverty and children', *Environment and Urbanization*, 14 (2): 23–43.

Newell, P. (2003) 'Towards a European Child Friendly Cities Initiative', unpublished conceptual paper developed for workshop at the UNICEF Innocenti Research Centre, 7–8 July.

Riggio, E. (2002) 'Child friendly cities: good governance in the best interests of the child', *Environment and Urbanization*, 14 (2): 45–58.

Satterthwaite, D. (1996) *The Scale and Nature of Urban Change in the South*, report for Human Settlements Programme, London: International Institute for Environment and Development.

Satterthwaite, D., Hart, R., Levy, C. *et al.* (1996) *The Environment for Children: Understanding and Acting on the Environment Hazards that Threaten Children and Their Parents*, London: UNICEF/Earthscan.

Spencer, C. and Woolley, H. (2000) 'Children and the city: a summary of recent environmental psychology research', *Child Care, Health and Development*, 26 (3): 181–98.

Stansfield, S., Brown, B. and Haines, M. (2000) 'Noise and health in the urban environment', in Buxhner, V. (ed.) *Reviews on Environmental Health*, London: Freund Publishing: 43–82.

Swart-Kruger, J. (2002) 'Children in South African squatter camp gain and lose a voice', in Chawla, L. (ed.) *Growing Up In an Urbanizing World*, London: UNESCO/Earthscan.

UN (1992) *Agenda 21: The Rio Declaration and Statement of Forest Principles*, New York: United Nations Publications.

UN (2003) *Improving the Lives of 100 Million Slum Dwellers*, New York: United Nations Publications.

UNCHS (1996) *An Urbanizing World: Global Report on Human Settlements 1996*, Oxford: Oxford University Press.

UNCHS (1997) *The Istanbul Declaration and the Habitat II Agenda*, Nairobi: UNCHS.

UNICEF (1990) *Plan of Action for Implementing the World Declaration on the Survival, Protection and Development of Children in the 1990s*, New York: UNICEF.

UNICEF (1992) *Convention on the Rights of the Child*, New York: United Nations Publications.

UNICEF (1996) *Towards Child Friendly Cities*, New York: UNICEF.

UNICEF (1997) *Children's Rights and Habitat: Working Towards Child friendly Cities*, New York: UNICEF.

UNICEF (2001a) *Partnerships to Create Child Friendly Cities: Programming for Child Rights with Local Authorities*, New York: UNICEF/IULA.

UNICEF (2001b) *Young Voices: Opinion Survey of Children and Young People in Europe and Central Asia*, Geneva: UNICEF.

UNICEF (2002a) *A World Fit for Children*, New York: UNICEF.

UNICEF (2002b) *Poverty and Exclusion Among Urban Children*, Florence: UNICEF Innocenti centre.

UNICEF (2004a) *Building Child Friendly Cities: A Framework for Action*, Florence: UNICEF Innocenti Research centre.

UNICEF (2004b) *Definition of a Child Friendly City*, UNICEF. Available online: www.childfriendlycities.org/resources/index_definition.html (accessed 24 October 2004).

UNICEF (2005) *State of the World's Children 2005*, New York: UNICEF.

UNICEF (undated) *How to Build a Child Friendly City*, UNICEF. Available online: www.childfriendlycities.org/resources/index_building_cfc.html (accessed 24 October 2004).

Chapter 3

Australia's toxic cities

Modernity's paradox?[1]

Brendan Gleeson

Not seen, not heard

There is much contemporary hand wringing over the implications of population ageing. Not a week passes, it seems, without a new report urging attention on the issues arising from an ageing Australia. But what about children? They seem to have disappeared from public focus, only to return in bouts of wild panic concerning abuse and various other calamities that occasionally touch young lives.

Governments and business lobbies may want to prepare us for a greyer population. The prospect must frighten neo-liberal ideologues: legions of baby boomers downshifting and escaping the treadmill. But are we planning for a future Australia that will nurture children and youth? Australia's population demography debates have led us to be obsessed about a very partial view of the future: one fixed on ageing baby boomer legions flocking to new coastal lifestyle regions. However, the future will contain a far greater range of human needs and interests than this simple picture would suggest.

For a start, Australia is, and will remain, a thoroughly urban society, with most of its citizens living in the suburbs of the principal metropolitan regions.[2] Australians overwhelmingly continue to prefer living in the sub-regions of the main cities. Australia's future will not be decided by the fortunes of new coastal 'sea change' regions, with their flourishing retirement communities, but by the ability of its suburbs to provide inclusive and sustainable living environments for the bulk of our population. And in many places this population will include more children, not fewer, than is presently the case. It is strange that this reality is so absent from public and popular debates.

The emergent will to brush children aside may reflect in part the cults of individualism and materialism that have flourished during the era of neo-liberal reform. Children necessarily restrain the lure of individualism – they are by nature dependent. They remind us that we are *of Nature*. Their upkeep dilutes the material flow and their care reduces the time for self. Every social prohibition has its geography and thus we will surely witness the surfacing of new islands of *demographic* exclusion, such as Aurora, the child-free 'community' proposed for Queensland's Gold Coast region by developer Craig Gore. Aurora was scuttled in 2003, after raising the ire of the state's Children's Commissioner, but its proposal raised a semaphore warning of the new archipelago of exclusion lying on the horizon of selfishness that we are steadily approaching.

Walls may keep the kids out but population ageing will not free us from their claims. An increasing weight of evidence tells us that a growing number of older Australians, typically grandparents, are assuming the role of primary carer for children. Nearly 28,000 children aged under 15 are presently being raised solely by grandparents (Passmore 2004). Many have had to step into the social breaches created by the storms of political economic change as parents succumb to mental and physical breakdown, drug dependency and criminality. Others are providing the care that working parents cannot access or afford. A mixture of love and necessity drives these older Australians back to parenting, often in stressful circumstances. An overwhelmingly urban nation must look to its cities to provide healthy living environments for our young and for the increasingly diverse range of people who care for them. There is reason to be gravely concerned about the steadily accumulating health and social commentary that is reporting deterioration in the life worlds of young Australians.

In contemporary Australia, great shifts are under way within children's home worlds and within the policy areas that structure their lives, especially education, health and childcare. Decades of neo-liberal reform by successive national and state governments have deepened social and geographic polarization and caused the contraction and deterioration of urban public domains (Forster 2004). These urban shifts threaten to reduce the life chances of the poor and their young, and to deepen their entrapment in degraded urban realms. The drift to segregated cities and the loss of civic realm also undermine the well-being of children from wealthier households and communities. In the new islands of suburban privilege, the relative absence of a public domain impoverishes the young in a different way, excluding them from the principal civic resources and social experiences that nourish the development of strong citizenship values.

This chapter explores some of the implications of urban structural change in Australia for children at both ends of the wealth scale. It charts the emergence of 'toxic cities': urban areas that fail to nurture the young, and that increasingly threaten them with physical and mental harm. The portrayal is a grim indictment of neo-liberal reform, which has taken us progressively further from the idea of child friendly cities. It begins by reviewing 'toxic' changes to the urban public realm, including transformations to childhood services that have exposed children to the vicissitudes of profit driven 'care'. The next section reviews evidence that testifies to the ways in which the culture of neo-liberal productivism and individualism has failed children, whose physical and mental health seem to be in decline. Many of these indicators of childhood morbidity have been evident in other Western, especially Anglophone, countries, and point to a deeply harmful paradox that afflicts the pursuit of wealth. The lustre of materialism is inevitably dimmed by its toxic consequences for children. The last two parts of the chapter point to the different, yet inevitably harmful, legacy of toxic cities for children at both ends of the wealth scale.

Malign care

Early childhood services have been a key theatre in the war on the public realm perpetrated through two decades of neo-liberal reform in Australia. In the past decade, this cottage, community-based industry has been transformed by corporatization, rising public subsidies and the rapid growth and scale of for-profit provision. In 1990, for example, about 80 per cent of Queensland's childcare centres were managed by the community sector – by 2005 the ratio was reversed, with corporate entities such as the ABC Learning Group, providing the bulk of care (Elvish 2005). Just over 70 per cent of Australia's 4,300 childcare centres are now run for profit (Murdoch 2004). During this transformation, quality community-based centres have struggled to compete with corporate providers, which emphasize lower wage costs and mass service provision. Brian Elvish, chief executive officer of one key community-based provider, C & K, remarks:

> the prime culprit in the rise of the corporate sector and the demise of the community sector is the Federal Government. Its free-for-all development policies have encouraged entrepreneurs by significantly increasing the number of childcare places available. However, the simultaneous removal of operational subsidies to the community sector has *virtually eradicated the right of choice* for . . . families.
>
> (Elvish 2005: 17, emphasis added)

He points to a fundamental difference in the quality of service provided by community-based centres, which emphasize parental accountability and individual child welfare, compared to that offered by the corporates. These, 'like the fast food industry', are dominated by the logic shareholder accountability and thus cost minimization (Elvish 2005). For Elvish, the 'politics of choice' are fraught with paradox: 'In our consumer orientated society we are bombarded with choice and variety. Yet ironically, at one of the most important periods in a child's life, families are increasingly given a "one size fits all" option to early childhood' (Elvish 2005).

The market-driven approach is also causing a significant spatial mismatch between the location of commercial centres and the geography of parental need. Thompson reports 'a chronic shortage of childcare places in many urban neighbourhoods where land is more expensive' (2005: 1). Consequently, '[p]arents are being forced to make direct choices: leaving their children unattended or in "backyard care" minded by untrained strangers or aged relatives' (Thompson 2005: 2). Windfall profits, driven by relentless cost minimization and by record federal subsidies, are not leading to pay improvements in a notoriously low-pay industry, or to better services in high-need communities. Flood's study demonstrates that low- and middle-income parents struggle to access formal care. He concludes: 'poorer parents are priced out of the care they need for their children, and parents in disadvantaged communities are more likely to find that no childcare places are available' (Flood 2004: 7).

The issue is hardly a social sideshow. The relentless parental enslavement to work, driven by decades of neo-liberal productivism and the consequent loss of family time (especially weekends) (Pocock and Clarke 2004) have rendered childcare one of the most important domains for socialization and a key determinant of child – later adult – well-being. There are now over 730,000 Australian children in formal childcare (up from approximately 570,000 in 1999) and an unknown number negotiating the vagaries of informal care. The shift in logic from a focus on children as individuals with specific care needs to 'units of subsidy' whose service costs must be minimized marks a transition to a heartless, and ultimately self defeating, mode of childcare.

For children, the problems arising from human service changes have been compounded by the public liability crisis that represents yet another solvent force eating away at the urban public realm. The early years of the new millennium witnessed an outpouring of concern from local government and community organizations, who pointed to a rapid collapse in the quality and volume of community cultural and sporting activities. The public liability pandemic is not, however, a 'natural' outcome reflecting the inherent vulnerability of children in urban contexts but arises simply from the unwillingness of governments, especially the federal government, to secure the health and safety of citizens in civic spaces. The storm of liability claims has broken most severely across the municipal public realm – most of Queensland's local governments, for example, now regularly have major (i.e. over A$5,000) liability claims made against them relating to incidents occurring in public places (King 2005).

The crisis might well reflect what Greg Hallam, Executive Director of the Queensland Local Government Association, describes as a growth in the 'absurdity and the avarice of people' (in King 2005: 5). He blames the long reign of the 'nanny state' for a culture of public dependency, but neglects the fact that Nanny, if ever she existed, has long departed the scene, ushered off to oblivion by the Prophets. Perhaps more accurately, the explosive growth of public liability litigation reflects the melding of three social forces that have been deliberately conjured during the neo-liberal reform phases: heightened individualism; fading respect for the public realm; and the relentless growth of economic insecurity. The latter anxiety manifests itself in public liability claims when individuals, frightened by the perceived or real decline of public support for health and human care, attempt to find other ways of registering their potentially reasonable claims for social assistance after the experience of some injury.

The net result is a rapid contraction in the size and quality of the public sphere that services children. The journalist Lawrence (2005a: 42) reports: '[p]ublic spaces are becoming boring because of public liability fears'. She cites the urban designer and scholar, Danny O'Hare, who observes that a culture of obsessive risk aversion, especially among local governments, has caused a widespread decline in the quality of public play environments for children. The guardians of these public sphere playgrounds cannot afford the hefty insurance levies that a more sanguine approach would attract. Lawrence's report details the outright removal of play equipment in many Queensland municipalities, and a shift away from adventurous and stimulating designs in other contexts. In 2002, a people's panel, convened by the Victorian Government, linked the public liability crisis to children's declining health, especially citing evidence of growing childhood obesity. The panel lamented the culture of risk that had blanketed the public realm, especially the reduction of the ability of public schools to act as centre points of child focused community activity (Victorian Government 2002).

Modernity's paradox: fatter, sicker and sadder

The growing endangerment represented by these and other shifts is surely reflected in accumulating scientific evidence that reports that our children are getting fatter, sicker and sadder. Australia has become a much wealthier country over the past three decades, but this material enrichment has been accompanied by a startling decline in the health and well-being of children. A similar pattern of simultaneously rising rates of wealth and morbidity has been registered in other developed countries – the Canadians Keating and Hertzman (1999) have described this as 'modernity's paradox'. In the US, the psychologist Myers believes that compared to the 1950s, contemporary Americans, 'are twice as rich and no happier. Meanwhile the divorce rate doubled. Teen suicide tripled . . . Depression rates have soared, especially among teens and young adults. I call this conjunction of material prosperity and social recession *the American paradox*' (in Luthar 2003: 1584, original emphasis).

Fiona Stanley, epidemiologist and 2003 Australian of the Year, reports an exhaustive review of physical and mental health indicators that shows that 'whilst death rates are low and life expectancy is terrific, *trends in almost all other outcomes [for children] have got worse*' (2003a: 2, emphasis added). Consider just some of the indicators that have registered declines for children during the era of neoliberal structural reform: decreased birth weight and increased post-neonatal mortality (Aboriginal children); asthma and diabetes; obesity; intellectual disability; depression; anxiety; behavioural problems; drug use; and child abuse.

The dismal picture was recently confirmed by an Australian Institute of Health and Welfare (AIHW 2005) study that reported rising obesity and mental health problems for Australian children, and indications of worsening dental health. It was not all bad news: Stanley's observation about improving infant mortality rates was confirmed, though for indigenous children the rate was still 2.5 times higher than that of others. This surely reflects the power of *public* health to improve individual life chances. One worrying indicator of child safety showed the number of children on care and protection orders to have risen almost 50 per cent in the past six years (AIHW 2005).

For youth, accumulating evidence points to a sadder, more fragile outlook than was previously the case. Again, not all trends are in decline. Eckersley points to 'fractured views' of youth well-being, including positive assessments based on self-reported life satisfaction studies against countervailing research on mental health that shows many young people are struggling. He concludes that these seemingly contradictory trends (also evident in other Western nations) are not necessarily irreconcilable and point to new complexities in the human condition. Most young people are, and have always been, optimistic about their own futures, but most are pessimistic about the state and prospects of society or the world. Most do not believe quality of life in Australia is improving. They are more likely to think that globally, this century will be a time of crisis and trouble than that it will be an age of peace and prosperity (Eckersley 2004: 38).

The fractured reality of youth well-being may reflect in a specific way the Great Uncertainty (Kelly 1994) that has come to afflict the Australian popular outlook after the prolonged structural reform and cultural pluralization that has taken place since the 1960s. Youths, perhaps more than any other social (certainly demographic) group, are best placed to positively embrace new opportunities and experiences. These have arisen through cultural pluralization and the new fluidity of employment structures and of life courses. And at the same time, they may be most vulnerable to the heavy mantle of uncertainty that has settled over life in general, coupled with the greatly increased social expectations that they relentlessly engage in – 'reflection, reinvention, and flexibility' (Eckersley 2004: 38). The transience and impermanence of liquid neo-liberal modernity pose heavy psychic costs for the children of Freedom. Eckersley enlists Bauman (2003), who insists that for every head in the currency of change there exists a tail: the 'exhilarating adventure turns into an exhausting chore' (in Eckersley 2004: 38).

Again, what is lost to youth, as to many, is Time – a humane temporality that allows for studied, socially supported growth and adjustment that contrasts with the constant, instant reaction to chronic change demanded of contemporary young people. The endless acts of instant gratification (or disappointment) that choreograph the life of *Homo consumens*, the ideal subject of neo-liberalism (Bauman 2003), are an existence marked by the absence of Time. If Time has been stolen, little wonder that youth report to surveyors little enthusiasm for the Future – an unfolding of Time over which they will have little mastery. The yearning for Time stolen, for Time never experienced, resonates in the social surveys that register children's loss. A recent study by The Australia Institute recorded the laments of children who desired more time from their parents, not more family wealth (Pocock and Clarke 2004). One youth, 'Smithy', aged 17, told researchers, 'I wouldn't mind seeing them [parents] make less money if it meant they don't have to work so damn hard' (in Horin 2004: 6). 'Kelsey', aged 12, grieves for parental time, '[d]ad earns money . . . but he hardly spends any time with you' (Horin 2004).

While poor youth, as with children, are most exposed to the brutish contemporary risks to well-being – notably, homelessness and violence – social evidence suggests that liquid neo-liberal modernity presents much more general threats to young Australians, especially to their psychic health. The growth in materialism and individualism that has accompanied both phases of the neo-liberal reform project – and most acutely the last – seems, as Eckersley observes, to 'breed not happiness but dissatisfaction, depression, anxiety, anger, isolation and alienation' (2004: 40). Young people with little historical experience of the many layers of social solidarity that have been peeled away during the reform project are most vulnerable to the chill winds of alienation.

The much-discussed growth of urban tribalism among Western youth (see Watters 2003) may be a positive self-correcting response to the gales of individualism and materialism, but their long term socializing potential is unknown. It may well be counterbalanced by anti-social forms of tribalism, such as the occasional outbreaks of disorder among middle-class youth that produced moral panics in the media, in parents and in governments in recent years. A case in point is the 'gatecrashing epidemic' that has surfaced in cities across Australia, from Perth to the Gold Coast, with the popular media raising the spectre of swarms of 'drunken often drugged-up gatecrashers' descending like youthful Visigoths without warning on homely family celebrations (Dibben 2005: 14). The anger and ennui unleashed in these events are hardly reflective of content, socially adjusted youth.

What happened to the great historical promise of modernity? Why has escalating wealth not also lifted the prospects for our young? How can it be that in the first years of the new millennium, two centuries after the Industrial Revolution, Australia's leading scholar of the young could declare: 'childhood is rapidly vanishing' (Stanley 2003a: 8)? Fiona Stanley writes:

> Clearly, our nation's economic prosperity has failed to deliver the social dividend that was promised. While Australia prospers economically, *alarm bells have been sounding in the suburbs* – witness increases in divorce, family violence, child abuse, homelessness, working hours and social isolation.
>
> (2003b: emphasis added)

Could it be that the centuries-long 'Growth Fetish' (Hamilton 2003) has produced cities and communities that are environmentally and socially injurious to their most vulnerable human inhabitants, the children and the poor? In explanation, Stanley speaks of the shattering consequences of the growth machine economy and the social changes engendered by this: unprecedented levels of family breakdown and discord; ever longer working hours; cultural alienation; and rising wealth inequality. The weakening and withdrawal of the public domain from many urban communities has surely left children vulnerable:

> What's been decreasing are some of the protective factors for these things [morbidity levels]: community cohesion and participation, neighbourhood trust, and I think, children's services and facilities in many communities . . . have you talked to any child health nurses lately to see how angry and worried they are about support? . . . there has been a decrease in facilities at a time when parents actually need them more than ever . . .
>
> (Stanley 2003a: 11)

Macro indicators mask how vulnerability affects young Australians in different ways. As noted above, wealth polarization – what Fiona Stanley terms the 'toxic social divide' – produces distinct forms of endangerment for the young. In our new urban poverty spaces, the endangerment is real, even life threatening.

The jaded and foul

One Saturday in early November 2003, the lifeless body of five-year-old Chloe Hoson was found discarded amidst refuse in the reserve opposite her home, Lansdowne Caravan Park, in south-western Sydney. Chloe had been raped, strangled and cast aside like rubbish by her killer. A young man resident in Lansdowne Park was later charged with her murder. In a moving and insightful piece of journalism, Christine Jackman, of *The Australian* newspaper, took readers beyond the monstrous crime that ended Chloe's life into the life world that she had inhabited all too briefly (Jackman 2003). Her essay recalls the higher forms of Victorian era slum journalism, which explored the realms of urban abjection and recorded them with humanity and empathy.

Jackman enters the Lansdowne netherworld to find and interview Chloe's playmates, just some of the many children who live with their parents in the tightly packed, poorly ventilated trailers and cabins that crowd the caravan park. The children knew well the tawdry reserve where Chloe had been found because their parents had declared it off limits: a magnet for prostitution and drug dealing where local council workers would regularly find up to 150 syringes a week buried in the sand beneath the playground's swings.

Jackman speaks to 'Dave', a father of three children under the age of three. He fumes with helpless rage about the impossibility of keeping kids safe in a sinkhole of drugs, pollution and quietly smouldering rage. Jackman observes: 'beneath this father's fury is a deeper, brooding resentment at the powerlessness of life on the fringes of Australia's wealthiest city.' Dave speaks of entrapment, of not being able to access even the scruffier private rental market that thrives in parts of Sydney's middle west. He and his family have been told that they will wait at least six years for public housing in the area. The heartless contradictions of contemporary post-welfare Australia are revealed when his partner, 'Cara', informs Jackman that the family could move almost immediately to a public dwelling in the city of Dubbo (in rural New South Wales (NSW)), but the move would send them even further backwards: 'Centrelink [the national government's welfare coordination agency] will cut our [unemployment] payments because you can't move to an area with less jobs.' The long shadows of Victorian poor laws and their brutish prosecution of the 'undeserving' continue to darken the lives of Australia's urban poor.

Lansdowne, like many other similar urban welfare camps, is the last stop before outright homelessness. It provides none of the conditions for a healthy and happy life. In Dave's words: '[t]here's nothing here, mate.' The politics of choice seem to have side-stepped Lansdowne's 1,000 residents: 'we've got no choice', laments Dave. Jackman writes: '[t]he only shopping centre within walking distance boasts a liquor store, a Chinese takeaway and a McDonald's – but the fruit and vegetable shop has closed down.' The problem of transport poverty is highlighted: '[t]hose without cars must rely on the local service stations for ready supplies – but often must dodge another sort of trade on their way to pick up milk.' Jackman inscribes her sad portrait of Chloe's life and death with this epitaph: '[s]he was an innocent battling to thrive in a world where the fresh and the natural are constantly under siege from the jaded and foul.'

Australia's cities are peppered with 'jaded and foul' places that are home to countless numbers of children and youth. More than one-fifth of the nation's children live in low-income households, and nearly as many live in families with no employed parent (AIHW 2005). Studies by Lloyd, Harding and Greenwell (2001) and Randolph and Holloway (2004) separately confirm that many of the poorest urban localities in Australia contain high concentrations of children (i.e. over 25 per cent of their population). These places include the new and lingering concentrations of poverty that have emerged in Australia's cities during decades of neo-liberal reform: welfare camps like Landsdowne; sinkhole public housing estates; places of urban indigenous disadvantage; and emerging pockets of exclusion mouldering in the middle suburbs of the major cities.

Glenn Draper and colleagues have studied the relationships between health inequalities and mortality, examining the influence of socio-economic disadvantage on life expectancy. Their national study of socio-economically disadvantaged communities confirms the argument that poverty is literally life threatening. They come to the chilling conclusion that, '[i]f all [poor communities] in Australia experienced the same death rate as the least socio-economically disadvantaged areas, more than 23,000 deaths could have been avoided in

1998–2000' (Draper, Turrell and Oldenburg 2004: 22). *This figure includes nearly 2,700 children.*[3]

Caravan parks like Landsdowne are among the worst and least-known communities of disadvantage. The little we do know about these new urban poverty spaces was powerfully summarized in a study by researchers at the Urban Frontiers Programme at the University of Western Sydney (Wensing, Holloway and Wood 2003). In 2001, 62 per cent of households in caravan parks earned less than A$500 a week, compared with an Australian average of 29 per cent. More than four in ten park residents were in rental stress, paying more than 30 per cent of their income on rent. Some 80 per cent of residents had no post-school qualifications. We do not properly know yet the extent or precise character of these new urban netherworlds. Our social scientific understanding of them is poor. Such knowledge seems to attract little political or policy interest.

Epilogue? The problem of Lansdowne, at least, might soon be dealt with. The owner of the caravan park, urban development behemoth Meriton, recently lodged an application to rezone the park from 'private recreation' to 'residential', potentially paving the way for a lucrative redevelopment. Meriton has announced that it is 'considering options for the best use of the land occupied by the park. At this time, it is considered that a traditional, low density, residential subdivision is the most appropriate use'.[4] A master planned enclave estate might be just what is needed to sanctify the haunted grounds of Lansdowne Caravan Park. But who will calm its ghosts?

Poor (not so) little rich kids

What about the other end of the social scale? Imagine the situation of the children and youth who might live in the master planned estate that could eventually replace Lansdowne caravan park. It is also likely to look very crowded in a middle-class way; lots of large houses packed into small lots, separated by narrow streets and pocket parks. It may or may not have footpaths.

Some Australian commentators have derisively called such estates 'McMansion Land' (see Figure 3.1). In truth, the observation is condescending and rather unfair: the large structures reflect a growth in the national appetite for more housing space that has been a feature of Australian life for much of the twentieth century and now continues beyond.[5] At the same time, the plots on which master planned housing estates are set have been dramatically compacted, and decades of 'urban consolidation' policies been applied to new subdivisions by state and local governments. These endless little compactions, and the general speeding up of life under neo-liberalism, are features of what Kearns and Collins in New Zealand (Chapter 7) term the 'intensifying city'.

Hawley, resorting to mild hyperbole, describes contemporary project homes on the newer Sydney master planned estates as 'four bedroom, spiral staircase,

3.1 Where's the space for play? Contemporary urban development in Brisbane, Queensland.
Source: Neil Sipe.

open-plan, kitchen-family-dining-lounge, multiple bathroom, rumpus room, big-screen media room, barbecue, spa, multi-garage bigger-is-beautiful-is-better houses' (2003: 25).

While condescension is unwise, there are growing reasons for disquiet about McMansion Land. The growth in housing girth is an environmental concern: the suburban palazzos are energy guzzlers – and also, perhaps, a health concern. Evidence on the national epidemic in childhood obesity points to a relationship between the expanding girth of dwellings and the growing waistlines of their inhabitants.

The contemporary suburban mega house internalizes activity, allocating large amounts of space to passive recreation: home theatres, lounges, rumpus and computer rooms, courtyards, and monster garages for the storage of adults' toys. Gwyther explains: 'They love cocooning inside their McMansions, which are like castles, fun factories and mini resorts in one' (in Hawley 2003: 25). These relatively sedentary residential landscapes contrast with older suburban forms that were premised on far greater levels of outdoor activity, especially for children:

> The traditional backyard has gone, along with its trees, garden veggie patch, often pool, washing line and shed, where children could let their bodies and imaginations run free and build tree houses, cubbyhouses, billycarts, dig in the dirt and invent

> games. Now, it's indoor computer games, and, given there's no room for a decent run-up in most McMansion courtyards, children are driven to sport and formally organised activities most days of the week.
>
> (Hawley 2003: 25)

Recent survey evidence confirms the picture of sedentary children. A national study reported in early 2005 found that Australian children were spending only about one-tenth of their time in outdoor play (Allen and Hammond 2005). Further, one in twenty admitted that they never left their homes to play. In response, the parenting educator Michael Grose linked rising childhood obesity to the fact that 'houses are getting bigger and backyards smaller' (Allen and Hammond 2005: 1). He also decried the programming of children by parents and educators, observing, '[e]verything we know about child development says that children need to spend more time outside in unstructured activities, giving them a chance to explore their environment' (Allen and Hammond 2005).

Those 'McKids' who actually do participate in organized sport – a chore for parents working long hours on the mortgage treadmill – will experience at least some level of physical activity. But missing from these new suburban landscapes are the opportunities for spontaneous, constant, free play that was available to children of previous generations, and that is available to those lucky enough to have backyards still. As Hawley observes, many parents cite space as the principal reason for rejecting 'inner city shoe boxes' in favour of the new master planned estates. And yet free, permeable space seems to be almost absent from the new residential landscapes.

The freedom and permeability of activity space is further reduced by the highly routinized and supervised lives imposed on contemporary middle-class urban children. The geographer Paul Tranter believes that Australian children are subjected to unprecedented levels of surveillance and control, driven by an epidemic of parental and institutional concern about environmental risk and crime (Chapter 8). Many now live highly scripted lives, marked by pervasive anxiety and the absence of free and independent play. Cadzow writes of the 'Bubble Wrap Generation':

> So reluctant are we to let our offspring out of our sight that we drive them to the playground and everywhere else rather than allow them to walk or ride their bikes. Strapped into the backseat of the family sedan, chauffeured to and from school, soccer practice and piano lessons, middle-class Australian boys and girls are like pampered prisoners – cosseted, constrained and constantly nagged.
>
> (2004: 18)

Children need autonomy from adults for their psychic and social development: little wonder then that the 'pampered prisoners' flee the bounds of their suburban cells for the horizonless expanses of computer-generated worlds, where free play is always possible. They may not be permitted to climb trees, ride their bikes to the shops or go unaccompanied to parks, but here they can wage global or intergalactic wars, build cities and even design the perfect family.

The problem with simulated worlds, however, is that they are pretty poor training grounds for life. A tour of duty in SimCity can never emulate the sensuous complexity of urban life. Life with the Sims is unlikely to help a child to cope with any family dysfunctionality or prepare them for the joys and strains of adult life. The 'ordinary maladies' of life come as insurmountable shocks to bubble wrap kids. Melbourne clinical psychologist Andrew Fuller tells Cadzow, 'when bad things do happen, they're just thrown for six. They end up in my bloody therapy room and I'm sick of it' (in Cadzow 2004: 19).

By the time they reach their teenage years, many children will be quaking from prolonged exposure to the chill winds of materialism and individualism. For many, drugs – illegal and prescribed – will help to stop the shakes. Health Insurance Commission data show that Australia has joined the ranks of Prozac nations, reflecting rapid growth in prescriptions for happy pills during the early years of the new millennium. In 2004, over a quarter of a million prescriptions were issued to adolescents to treat various forms of sadness (Lawrence 2005b). Professor Ian Hickie, mental health expert, warns of 'the future burden of youth mental illness' arising from a rapid growth in conduct and mood disorders among young people (Lawrence 2005b: 39).

Many of the saddest children are from affluent homes. Australian and international evidence confirms that children frequently bear what the US psychologist Suniya Luthar calls the 'psychological costs of material wealth' (Luthar 2003; in Australia, Eckersley 2004). Empirically based research links disorders that appear to be growing among wealthier youth – notably, substance abuse, depression and anxiety – to a combination of pressure to achieve and isolation from parents. She crafts the image of the remote yet overbearing parent. Luthar reports a variety of US studies that suggest that the costs of wealth are especially acute in upwardly mobile suburban communities. In Australia these costs are apparent in the aspirational belt, where pressures to excel in academic and extra-curricular activities combine with isolation from parents working long hours in the contemporary treadmill economy (Luthar, in Eckersley 2004).

Despite evidence that shows Australia to be a much safer place for children than it was three decades ago,[6] an obdurate culture of fear drives the ever increasing parental colonization of children's life worlds. The colonization project seems strangely disconnected to real social evidence. It includes, for example, a recent and largely unreported Australian Bureau of Statistics survey that shows a significant drop in crime in New South Wales between 2001 and 2003, and an increase in the number of people who reported that their neighbourhoods were crime free (Totaro 2003). Parental anxiety about child abduction, driven by relentless, high-pitched media reportage, means that many more children are driven to school than previously. One result is growing traffic chaos around schools, which perversely poses new dangers for children. The president of a parents' association at a Gold Coast school recently remarked:

> It's absolutely hideous. There's [*sic*] cars everywhere. It's dangerous around schools and it's all because of the child abduction issue . . . It's a bit sad. Kids don't have much freedom . . . They can't be kids any more. We feel like we are part time security guards. It's the way society has made us these days.
>
> (in Robson, Weston and Lawrence 2005: 4)

Fear, the potent flag bearer for Despair, is a difficult wraith to banish, especially when popular media and contemporary 'law and order' politics give it the free pedestal that it so desperately craves. The rising numbers of kids in therapy and the epidemic of child obesity are two potent markers of the extent to which fear rules our cities and communities.

Postscript: a battle quietly rages

Like most advanced capitalist nations, Australia has long craved greater wealth, more freedom to use it and more stimulating ways to expend it. Before the neoliberal revolution three decades ago, the lust for gold was restrained by a diverse set of moderating influences with deep cultural roots: conservatism, religion, socialism, conservationism. The manipulated panic about the 'state fiscal crisis' that brought neo-liberalism to power throughout the English speaking world in the late 1970s saw the suspension of these moderating orders. The 'growth machine' economy unleashed by neo-liberals promised to drive whole nations to heaven through the eye of the materialist needle.

The assessment presented in this chapter suggests that the Growth Fetish is a morbid desire, and its indulgence has generated material enrichment at the cost of great civic and human impoverishment. From the perspective of children, it has produced toxic, not healthy, cities. A nation that denies the chance for health and happiness to many of its young is not a rich society, because it is brutish by nature and because it steals from its own future. The attack on young people suggests that we have, as a society, lost the (re)generative impulse that is a precondition for a national future worth having.

Further, the open disregard of successive national and state governments and business elites for the highly apparent, blatant polarization of cities reeks of doom. The steady, nihilistic progress towards an Australia divided is surely a death march. A splintered nation will not weather history's storms. Finally, the attack on the public domain represents another way in which Australia has campaigned against its own future. It has produced urban communities and cities that cannot undertake the task of nurturing and constantly modernizing the civic values that underscore nationhood.

Amidst this ruckus, ordinary things are happening that will shape the future. Children and youth are trying to live and grow in the shadows of great events. The glowing structures erected around them, and the riches piled up in their sight, provide no shelter from human perfidy. Our misshapen young will pay the debts that we accumulate. Fatter, sicker and sadder, they face the miserable future we are shaping for them.

Notes

1 This chapter is reproduced in amended form from my 2006 book, *Australian Heartlands: Making Space for Hope in the Suburbs*, Sydney: Allen & Unwin.

2 Capital cities presently contain almost two-thirds (64 per cent) of Australia's population and their growth accounted for almost three-quarters of Australia's population increase in the 2000–1 period. Between 1996 and 2001, the capital cities collectively grew by 1.3 per cent while the balance of the States and Territories grew by 0.9 per cent (ABS (Australian Bureau of Statistics) 2003).

3 Aged 14 years or less.

4 Letter cited by Paul Lynch MP, state member for Liverpool, New South Wales Legislative Assembly, Hansard Article No. 32 of 1 May 2003.

5 The average floor area of new houses in Australia grew by over 40 per cent between 1984–5 and 2002–3. In New South Wales and Queensland, growth for the same period was over 50 per cent (Perinotto and Murphy 2004).

6 Cadzow (2004) presents Australian Bureau of Statistics data that demonstrate a marked decline in the number of deaths of children aged 5–14 between 1972 and 2002.

References

ABS (Australian Bureau of Statistics) (2003) *3218.0 Regional Population Growth, Australia and New Zealand*, Canberra: ABS.

AIHW (Australian Institute of Health and Welfare) (2005) *A Picture of Australia's Children*, AIHW Cat. No. PHE 58, Canberra: AIHW.

Allen, E. and Hammond, P. (2005) 'Modern kids prefer the great indoors', *The Courier Mail* (Brisbane), 14 February: 1.

Bauman, Z. (2003) *Liquid Love*, Cambridge: Polity Press.

Cadzow, J. (2004) 'The bubble-wrap generation', *The Sydney Morning Herald*, Good Weekend supplement, 17 January: 18–21.

Dibben, K. (2005) 'Bid to target party parents – fines move to curb gatecrashing epidemic', *Sunday Mail*, 22 May.

Draper, G., Turrell, G. and Oldenburg, B. (2004) *Health Inequalities in Australia: Mortality*, Health Inequalities Monitoring Series No. 1, AIHW Cat. No. PHE 55, Canberra: Queensland University of Technology and the Australian Institute of Health and Welfare.

Eckersley, R. (2004) 'Separate selves, tribal ties, and other stories', *Family Matters*, Winter, Australian Institute of Family Studies: no. 68: 36–42.

Elvish, B. (2005) 'Freedom of choice lost in childcare maze', *The Courier Mail*, 25 January: 17.

Flood, M. (2004) *Lost Children: Condemning Children to Long Term Disadvantage*, Discussion Paper no.64, Canberra: The Australia Institute.

Forster, C. (2004) *Australian Cities: Continuity and Change*, 3rd edition, Melbourne: Oxford University Press.

Hamilton, C. (2003) *Growth Fetish*, Sydney: Allen & Unwin.

Hawley, J. (2003) 'Be it ever so humungous', *The Sydney Morning Herald*, Good Weekend supplement, 23 August: 24–31.

Horin, A. (2004) 'Now it's generation T for time – with mum and dad', *The Sydney Morning Herald*, 7–8 February: 1 and 6.

Jackman, C. (2003) 'Chloe a victim of life on the fringe', *The Australian*, 15 November: 1.

Keating, D. P. and Hertzman, C. (1999) 'Modernity's paradox', in Keating, D. P. and Hertzman, C. (eds) *Developmental Health and the Wealth of Nations: Social, Biological, and Educational Dynamics*, New York: Guilford Press: 1–18.

Kelly, P. (1994) *The End of Certainty: The Story of the 1980s*, Sydney: Allen & Unwin.

King, M. (2005) 'Councils under attack from injury claims', *The Courier Mail*, 5–6 March: 5.

Lawrence, J. (2005a) 'Sorry kids, but fun is banned', *The Sunday Mail* (Queensland), 27 February: 42.

Lawrence, J. (2005b) '"Happy pills" crisis', *The Sunday Mail* (Queensland), 6 February: 39.

Lloyd, R., Harding, A. and Greenwell, H. (2001) *Worlds Apart: Postcodes with the Highest and Lowest Poverty Rates in Today's Australia*', paper presented to National Social Policy Conference, Sydney, July.

Luthar, S. (2003) 'The culture of affluence: psychological costs of material wealth', *Child Development*, 74 (6): 1581–93.

Murdoch, S. (2004) 'Subsidies turn childcare barons into millionaires', *The Courier Mail*, 2 October: 31.

Passmore, D. (2004) 'Grandparents to the rescue', *The Sunday Mail* (Queensland), 18 July: 56–7.

Perinotto, T. and Murphy, C. (2004) 'Carr homes in on McMansions', *Australian Financial Review*, 7–8 February: 15.

Pocock, B. and Clarke, J. (2004) *Can't Buy Me Love: Young Australians' Views on Parental Work, Time, Guilt and Their Own Consumption*, Discussion Paper no. 61, Canberra: The Australia Institute.

Randolph, B. and Holloway, D. (2004) 'The suburbanization of disadvantage in Sydney: new problems, new policies', *Opolis*, 1 (1): 49–65.

Robson, L., Weston, P. and Lawrence, J. (2005) 'Parents who must be security guards', *The Sunday Mail* (Queensland), 6 February: 4–5.

Stanley, F. (2003a) 'Before the bough breaks: children in contemporary Australia', Kenneth Myer Lecture 24 July, National Library of Australia, Canberra. Available online: www.nla.gov.au/events/seminars/kmyer03.html (accessed 30 August 2005).

Stanley, F. (2003b) 'Help young before it's too late: the early childhood agenda', *The Australian*, 21 November.

Thompson, T. (2005) 'Childcare sites fail to meet real needs', *Courier Mail*, 24 January: 1–2.

Totaro, P. (2003) 'Car and house thieves state's new endangered species', *Sydney Morning Herald*, 6 December: 3.

Victorian Government (2002) *A Healthy Balance: Victorians Respond to Obesity*, Melbourne: Department of Human Services.

Watters, E. (2003) *Urban Tribes: Are Friends the New Family?*, New York: Bloomsbury.

Wensing, E., Holloway, D. and Wood, M. (2003) *On the Margins? Housing Risk Among Caravan Park Residents*, Melbourne: Australian Housing and Urban Research Institute.

Chapter 4

Cities for angry young people?

From exclusion and inclusion to engagement in urban policy

Kurt Iveson

Introduction: child friendly cities and social control

Pressures to *exclude* so-called 'anti-social' young people from urban spaces continue to grow. Around about election time in countless jurisdictions in Australia and elsewhere, politicians of all persuasions can be relied upon to talk up concerns about urban disorder, incivility and crime. Promises are made about boosting police numbers and establishing tougher police powers to crack down on youths (and other potential 'enemies within') on behalf of 'communities' who are said to be threatened by their behaviour – all egged on by an expectant media. The fantasy of a city free of conflict is spatialized as a city that is safe for 'our children' and free of troublesome 'youths'. Of course, these exclusionary pressures meet with resistance. A range of organizations and individuals have challenged attempts to exclude young people from the city. Such efforts are premised on the belief that the exclusionary measures that dominate contemporary urban policy and politics are a major impediment to the creation of child- and youth-friendly cities.

But what are the alternatives to exclusion? Conventionally, critics of exclusion have argued for the *inclusion* of young people in spaces and processes from which they have been excluded. Progressive politicians, youth advocacy groups and many others have called for more inclusionary planning, inclusionary design, inclusionary policing and inclusionary political debate as correctives to the exclusionary impulses that seem to characterize so many urban spaces. However, I think the very notion of 'inclusion' has significant limitations. While I am certainly inclined to be more sympathetic to inclusionary efforts, I nonetheless believe that the exclusion/inclusion distinction provides a problematic framework for looking at young people's access to cities. Indeed, in this chapter, I question the usefulness of opposing exclusion with inclusion. It seems to me that exclusionary and inclusionary perspectives have more in common than we might at first think. To the extent that they are alternatives, they are too often only alternative routes to an agreed destination – a city that is orderly, conflict-free and controllable. In short, both exclusionary and inclusionary responses to the 'problem' of young people in cities are frequently mobilized on behalf of a vision of *social control*, in which difference is reduced to deviance, conflict is reduced to disorder, and 'the community' is reduced to 'the compliant'.

If this is so, then an urgent question presents itself for those who want to work for child friendly cities: can we conjure up a vision of child friendly cities that is not at its heart a vision of social control? This question guides the current chapter. I argue that we must find a way around the exclusion/inclusion distinction to conjure such a vision. Fortunately, some extraordinary activists and community workers are showing the way, working towards a kind of *engagement* with young people that starts from a different vision of the city and a different understanding of the 'problems' to be addressed by urban policy. In what follows, I elaborate upon this notion of engagement as an alternative to social control.

The chapter proceeds in three steps. First, I offer some general observations about social control in cities. Here, I argue that in recent years urban governance has shifted to a 'neo-liberal' framework. This neo-liberal framework has worked to construct 'circuits of exclusion' and 'circuits of inclusion' in our cities in pursuit of a new kind of social control agenda. Second, I consider how these neo-liberal social control agendas figure the 'child', the 'anti-social youth', and the 'responsible parent'. Here, I argue that these three figures are understood both as a problem for urban governance and as an incitement for neo-liberal forms of urban social control. The 'child' is understood as a future citizen in need of community protection, the anti-social 'youth' as the anti-citizen from whom the community needs protection, and the 'responsible parent' as the citizen who takes responsibility for the ongoing reproduction of shared community norms and values. Third, I present two instances of neo-liberal social control in action, focusing in particular on the ways in which the figure of the 'anti-social youth' has been enmeshed in circuits of exclusion and inclusion. In debates about the place of young people in Perth's city centre, and in debates about illegal graffiti in Sydney, there have been organized efforts to actively exclude particular groups of young people from public space. After surveying these efforts, I consider some of the inclusionary alternatives that have been proposed. In both of these cases, I show how some efforts to include young people fit comfortably within neo-liberal logics of urban governance. But in both of these cases, we can also see evidence of different ways of working with young people – a mode of engagement that does not require young people to stop being angry about their place in the (adult) world in order to be valued but rather seeks to engage with young people in building a shared project of social change.

Circuits of exclusion and inclusion: neo-liberal cities and social control

Fantasies of rational, orderly, controllable cities are of course not new – they have been with us as long as cities themselves. But there is growing evidence that a new kind of vision of urban order has begun to emerge in cities across Australia, New Zealand, Europe and North America in the latter decades of the twentieth century. This new vision is associated with an approach to urban governance that has come to be described as 'neo-liberalism' or 'advanced liberalism'. While the literature on neo-liberal urban governance is increasingly focused on identifying the particular forms different 'neo-liberalisms' take across time and space (see Peck and Tickell 2002), it is nonetheless possible to sketch out some shared characteristics.

In many areas of urban and social policy, we have seen the introduction of privatization, marketization and consumerization, with governments 'steering' and regulating rather than 'rowing' and providing (Rose 2000b: 324). All this is informed by scepticism about the capacity of political action through the state to bring about the good of individuals and communities. The best regulatory mechanism for social activity is thought to be the market, which encourages autonomous actors to size up the available information and make informed choices about what is good for them. The state has been reconceived as:

> [M]erely one partner in government, facilitating, enabling, stimulating, shaping, inciting the self-governing activities of a multitude of dispersed entities – associations, firms, communities, individuals – who would take onto themselves many of the powers, and the responsibilities previously annexed by 'the state'.
>
> (Rose 2000a: 96)

Individuals, families, organizations, communities and entire cities are said to know what is best for themselves, not government – they must find a way to be autonomous in a competitive world, and they must take responsibility for pursuing their own best interests. Of course, this does *not* mean the complete withdrawal of 'the state', but a recasting of its role. The state simultaneously 'rolls back' its involvement from some areas of social provision, and 'rolls out' a set of incentives and assistance packages designed to encourage the formation of these autonomous and responsible citizens who are capable of governing themselves by making the right choices in the marketplace (Peck and Tickell 2002).

Newly emerging measures in the domain of crime and social control are firmly enmeshed in this neo-liberal policy agenda. The principles should be familiar to us by now. Social control and urban order are to be achieved by urging and enabling citizens and communities to 'take upon themselves the responsibility for the security of their property and their persons' (Rose 2000b: 327). Government through *responsibility* is as dominant here as in other dimensions of social and economic policy. How is this to be achieved? According to Mitchell Dean (2002), government in the service of good order and social control becomes a matter of both *facilitation* and *authoritarianism*. This involves:

- assisting those individuals and associations who are willing to take responsibility for their own security and the civility of their community through partnership and the provision of 'opportunity';
- punishing those who refuse to take responsibility for themselves, who have rejected the opportunities provided for them to take their place in the community and thereby constitute a threat to civility and good order.

Thus, contemporary neo-liberal social control strategies:

> [C]an be broadly divided into two families: those that seek to regulate conduct by enmeshing individuals within *circuits of inclusion* and those that seek to act upon pathologies through managing a different set of circuits, *circuits of exclusion*.
>
> (Rose 2000b: 324, emphasis added)

Discourses of *community* are central to the establishment of these circuits of inclusion and exclusion – its ties, bonds and values become the *means* of government (Rose 2000b: 329). As Stanley Cohen (1985: 116) noted some years ago '[i]t would be difficult to exaggerate how this ideology – or more accurately, this single word – has come to dominate Western crime-control discourse in the last few decades'. Through the work of these agencies and the application of these technologies, the circuits of inclusion and exclusion start to actually map onto the spaces of the city itself. Those in the circuits of inclusion move through an archipelago of safe spaces that are secured from the 'anti-social'. The anti-social, on the other hand, are banished to peripheral spaces, or have their movements restricted through the imposition of 'good behaviour contracts' and incarceration (in a privately managed facility naturally).

Of course, in order for these agencies to act, they must be able to distinguish between those who should be included and those who should be excluded. This is a complicated matter. Working through this logic of control through inclusion and exclusion, a number of different groups can be distinguished:

A those individuals who have attained the capacity for autonomy in the neo-liberal circuits of inclusion and do not require assistance;

B those who need some small measure of assistance to maintain their capacity for autonomy in the circuits of inclusion because they are in some short-term bother;

C those who are potentially capable of being autonomous participants in circuits of inclusion but need to be trained in the habits and capacities required (and, we might add, protected in the meantime);

D those who are not competent to exercise their own autonomy for one reason or another (even though they have reached a mature age);

E those who are wilfully disruptive and get in the way of the establishment and maintenance of good order and must therefore be removed from the circuits of inclusion to the circuits of exclusion (adapted from Dean 2002: 48).

The practices of social control in neo-liberal cities take shape precisely in identifying and dealing with such different categories of citizens.

Neo-liberalism's children

It should not be surprising, then, that the 'child', the 'youth' and the 'parent' are key figures through which social and political concerns about civility, order and social control in urban life are articulated. In the figure of the 'child' is condensed all that is vulnerable and most at risk in urban society, all that requires intervention and protection. The child is the future citizen, an individual who cannot yet be responsible for themselves and is therefore in need of the community's protection. Children must be kept safe and exposed to opportunities to develop their capacities to become autonomous, responsible adult members of 'the community'. And so, with reference to the child, the identification of risk and threat – threatening situations, threatening environments, threatening people – is prioritized, and interventions are legitimated.

Parents are of course assigned a key responsibility here. They must choose which school is best for their child as education is marketized. They must expose their children to purposeful, constructive activities that will develop their talents and capacities. They must take all responsible steps to keep their children out of risky situations by keeping them off the street and off the internet – the two key spaces where popular discourse currently locates the monstrous strangers they might encounter. And where this cannot be prevented, they must teach their children how to identify and deal with stranger danger.

Some of the consequences of these approaches to urban governance to address the problem of the 'child' are the focus of other chapters in this collection. Suffice it to say that, as researchers such as Gill Valentine (1997) and Claire Freeman (Chapter 5) have shown, pressures on parents to minimize risk to their children through the exercise of parental responsibility have tended to restrict the spatial ranges of children. The city is experienced by children, perhaps even more than by adults, as an archipelago of bounded activity spaces to which they are taken, or for which they have been granted special 'licences' by their parents. Of course, the city is not totally 'locked down', and parents and children have both founds ways around such restrictions.

In contrast to those governance mechanisms enacted on behalf of 'the child' are those directed towards the figure of the anti-social 'youth'. Not so much a future citizen as an 'anti-citizen', these 'youths' constitute a threat that has to be neutralized (see Rose 2000b: 330). Here, it is the community that needs protection from its anti-social youth. As former New South Wales Premier Bob Carr infamously put it in his 1995 election platform: 'We simply can't allow roaming groups or gangs of youths, their baseball caps turned back-to-front, to stop citizens walking the streets, shopping or using public transport' (Australian Labor Party (NSW) 1994). Some efforts might be made to 'reintegrate' such youths into the community (perhaps through tender care, or perhaps through shaming – see Braithwaite 1989), but the community's patience has its limits: 'three strikes and you're out'. As Karen Malone argues (Chapter 2), young people on the streets are frequently viewed with either suspicion or pity.

And here again, parental responsibility is a key factor. Anti-social behaviour is seen to signal a failure in the socialization process for which parents must take responsibility. Remedial classes in parenting skills are offered when difficult kids are first identified by a variety of agencies. Crime control measures identified by Bob Carr include family assistance packages, because family problems produce 'children who won't or can't go home, and end up roaming the streets at night' (Carr 2002). And for those parents who can't or won't make the most of such opportunities to learn how to take better responsibility for their children, parental responsibility for the behaviour of anti-social youth is legislated.[1]

Exclusion, inclusion and engagement: two cases

Now I want to offer two brief snapshots of neo-liberal approaches to urban governance and their consequences for young people. First I will focus on attempts to regulate a particular place (the city centre of Perth) and then on attempts to regulate a particular practice (graffiti). In both of these cases, I aim to demonstrate how exclusionary and inclusionary forms of social control competed for ascendancy as responses to the 'problem' of 'anti-social youth' and public order. In each of these conflicts, I also offer evidence of a third model of working with 'youth' that seems to me to offer an alternative to these approaches – a model we might call engagement.[2]

Dealing with 'anti-social behaviour' in Perth

Perth has become the first city in Australia to formally decree a curfew for young people, following a trend that is already well under way in cities in North America and beyond (Collins and Kearns 2001). The Young People in Northbridge Policy instructs police to use their powers under child welfare legislation to remove children aged under 13 from the 'adult entertainment precinct' of Northbridge after dark, to remove children aged between 13 and 15 after 10 p.m., and to subject all others aged under 18 to a more 'hard line' approach. Launching the policy, Premier Geoff Gallop (2003) said:

> This is about protecting children who, quite frankly, should not be wandering the streets at night. . . . It is also about protecting the rights of people to go about their business in Northbridge without being harassed by gangs of juveniles.

In its first year of operation, 961 young people were removed from the streets of Northbridge. A staggering 88 per cent of those young people were Aboriginal. In response to persistent criticism, Gallop once replied:

> There are some leaders in our community who have expressed the point of view that we don't have any right to take these kids off the street. While you get some people saying that kids have a right to be there, that's sending a very, very bad signal out and it's making it harder for the Government to do what it should do on behalf of the community.
>
> (Manton and Magil 2003: 3)

The Northbridge curfew is but the latest weapon mobilized by government in a sustained campaign to reduce young people's presence in Perth's public spaces, particularly at night, on behalf of 'the community'. This campaign has been waged by governments of both stripes since the early 1990s, when Perth City Council, working together with the State Government of Western Australia, engaged in an extraordinary and expensive drive to redesign and regenerate the public spaces of the city centre. The political rhetoric had all the hallmarks of the neo-liberal approach to urban governance. The State Government articulated the need for the city centre to be revitalized in order to lift the profile of Perth globally, regionally and locally, in relation to its rapidly expanding suburbs (City of Perth 1995). There was plenty of talk about making improvements to quality of life by clamping down on crime and disorder. Such actions were to be taken on behalf of a vision of community that was constantly articulated to include workers, residents, shoppers and tourists. This vision of an included community took shape through references to those whose activities threatened the urban renaissance: beggars and the homeless, drug users, mental health patients, anti-social young people, etc. For these groups, *circuits of exclusion* were established.

These circuits of exclusion were constructed using a number of technologies and strategies, many of which are now familiar to observers of contemporary urban governance in the post-industrial West. First, new technologies of surveillance were introduced into key sites of urban restructuring. Perth was the first capital city in Australia to install a comprehensive network of CCTV (closed-circuit television) surveillance in its public spaces (see Figure 4.1). Second, the Western Australia (WA) police conducted a series of operations designed to clear the inner city of undesirables. The police JAG team (it stands for Juvenile Aid Group) used their powers under child welfare legislation to literally sweep the streets clean of young people not accompanied by their parents.

These powers were used because police were not targeting criminal behaviour as such but the much more nebulous category of behaviour now commonly referred to as 'anti-social'. Policing operations with names such as 'Operation Sweep' and 'Operation Family Values' were conducted with the intention of making 'the streets of the city and Northbridge safe for families' (McNamara 1994a: 9). Third, measures were introduced to encourage 'responsible parenting' and 'family values'. On returning children to parents after their sweeps of the city, police provided pamphlets and advice about parents' legal rights to use reasonable force to 'correct' their children's behaviour (McNamara 1994a: 9). 'Responsible Parenting Orders' designed to enforce parents' responsibility to prevent children from 'skipping school or engaging in anti-social or criminal behaviour' are currently on the legislative agenda of the Labour Government to bolster this dimension of urban governance (Office of Crime Prevention 2004). Fourth, a series of urban design measures have been implemented with a view to 'target hardening'. Public seating has been removed from areas where 'anti-social' young people are known to congregate. The public transport system has

4.1 Surveillance camera, Perth.
Source: Kurt Iveson.

been a particular focus for target hardening activity, as a means of cutting the flow of young people off at its (suburban) source. While in Opposition, Labour leader Geoff Gallop said in 1997 that passengers were being 'terrorized by teenage louts' (Crawford 1997: 7), and as Premier he committed an extra A$7 million to electronic surveillance and transit police in the 2003 State Budget (Government of Western Australia 2003: 14).

Now some cracks certainly appeared in this agenda. In particular, while business owners in the city centre were often the most vociferous supporters of the exclusionary measures discussed above, they were also concerned with the negative image of the city centre that these exclusionary measures produced. Constant conflicts between young people and urban authorities painted a picture of the inner city as a dangerous battleground, and this picture was potentially bad for business. So, for instance, the Northbridge Business Association expressed its satisfaction with police conduct during Operation Sweep, claiming that the operation had contributed to a 35-per-cent reduction in local crime. One of their number, however – a more sensitive soul – worried that such operations would not provide a permanent solution to the problems caused by 'gangs' of young people: 'We can't just put them in a truck and ship them out to the bush, because they will come back and come back more angry, with more resentment' (McNamara 1994b: 45).

Such concerns lent further support to those who advocated a more 'inclusionary' approach to the problem of social control in the newly restructured inner city. A more sustained effort to find ways to include young people in the life of the inner city, it was argued, would lead to the creation of a more 'harmonious city'. The report of the City of Perth's Youth Forum noted that young people were never going to be completely swept out of the city centre:

> [Y]oung people are mainly attracted to the city and Northbridge for social reasons and for the atmosphere it provides. This includes mainstream, ethnic, Aboriginal and at risk groups. It is therefore important to recognize that despite a lack of affordable things to do, the city has an atmosphere and special quality that will continue to attract young people.
>
> (City of Perth 1997: 43)

The authors claimed that: 'Attempts to exclude youth from these areas have generated anger, resentment, alienation and in some cases retaliation by young people. This harassment is most acutely directed at, and felt by, Aboriginal and "at risk" youth' (City of Perth 1997: 9). The report argued for a range of policies and facilities that would work against this resentment by securing a place for young people in the city centre. These included an improvement of late-night public transport, the provision of basic facilities such as public toilets (that would not require young people to gain entry to commercial premises), and more low-cost recreational opportunities. Such facilities, it was hoped, would also address young people's concerns for *their own security* – as the report noted: 'There is a paradox in that many young people see the need for more security, yet perceive police and security guards as "against rather than for youth"' (City of Perth 1997: 10). Advocates of the 'harmonious city' argued that any new facilities for young people would have to be located in places already frequented by young people in order to be successful – rather than stuck in some out-of-the-way peripheral space as is often the case with youth facilities informed by the 'out of sight, out of mind' approach. To enable this, the report mapped the 'youth cruising zone' of Perth's city centre, offering a pictorial representation of the places where young people liked to congregate.

The attempt to establish circuits of inclusion for a more 'harmonious city' was pursued through a range of government advisory bodies, planning committees and, to a lesser extent, the mainstream media. However, one of the problems faced by youth advocates was that access to these institutional spaces seemed to be premised on a shared understanding of the 'problem' to be addressed. As Howard Sercombe (1995: 89) noted in 1995 in relation to participation in media debates about young people and the city centre, 'it is unthinkable that any commentator could say that they were not concerned about Aboriginal juvenile crime', even if they disagreed about how it was to be 'solved'. In this context, attempts to secure a place for young people in the city centre, even when successful, could quite easily be put to work in containing and controlling young people's activities and behaviour within the dominant logic of neo-liberal urban governance. So, for instance, when youth advocates were successful in establishing the Shades of Black youth club night in Northbridge, run by the Nyoongar Alcohol and Substance Abuse Service (Prior 1997: 11), they could rightly claim to be addressing needs identified by young people through the 1997 Youth Forum and working towards a more 'harmonious city'. Nonetheless, the provision of a youth-specific recreation space for 'at risk' young people also had the added bonus for police of making young people's presence in the city more governable because it was more knowable. And so one night on their regular patrols of Northbridge, the JAG Team found two young white girls on their way into the club and stopped them entering, presumably 'for their own good' (Prior 1997: 11).

Of course, not everyone accepted the dominant framing of Perth city centre's 'problem' as one of 'anti-social behaviour' and juvenile crime. Nor did everyone necessarily prioritize the 'harmonious city' in their approach to working with young people. For others, the main problem to be addressed was the marginalization of some young people and the lack of opportunities for them to express their own point of view in an uncensored fashion. *GiBBER*, a magazine produced for young people in Perth, is a case in point. The production of *GiBBER* was coordinated by two youth workers, who sought contributions in the form of writing or artwork from young people between the ages of 12 and 18 'who don't usually have their say. This could mean anything from not being at school to being homeless or in Longmore [juvenile detention centre]' (*GiBBER* 1995: 1). It was *GiBBER*'s policy to provide a space in which excluded young people could actually express and represent themselves. The workers engaged in outreach to alleyways, leisure centres, detention centres and any other spaces where they thought they could find their target contributors and audience.

GiBBER published all sorts of material – from angry poetry to stories of pets and love and friends lost to the street, to artwork and discussions of drugs and alcohol use. Reading through the old editions of *GiBBER* gives one a remarkable window onto the ideas, perspectives and desires of some young people that is too rarely seen. From my perspective, it differs from attempts to 'include' young people in that it actually provides opportunities for young people to articulate their own visions of what the city is like and what it could be like, rather than seeking ways to address 'problems' that have been determined in advance and by someone else. Nor was participation in *GiBBER* premised on adopting any particular perspective, least of all one that fits within a vision of the 'harmonious city'. Some of the contributions are offered in a spirit of unchecked anger and frustration about young people's experience of the city. Rather than working for social control, the editors of *GiBBER* pursued a kind of *engagement* with young people that sought to build a shared project in which young people themselves discussed the 'problems' that needed to be addressed from their own perspective.

Unfortunately, this did not make the magazine popular with government agencies. The editorial of the ninth issue of *GiBBER* began by telling readers that the editors had been contacted by the Juvenile Justice Department of WA regarding the magazine's content. The department threatened to revoke the workers' access to detention centres unless the contents of the magazine were toned down. When the workers refused, they received a letter from the Manager of Education at the Ministry of Justice, which read in part:

> [A]s a government body entrusted with the care and welfare of the young people in their charge, Juvenile Justice is open to serious criticism should it continue contributing to and receiving the magazine. While GiBBER's print policy of encouraging alienated youth to express publicly their innermost thoughts and emotions has merit, it is a teacher's responsibility to manage that expression in terms that are community appropriate and acceptable.
>
> (quoted in Jefferys no date)

The ninth issue of *GiBBER* was the last. If ever we needed an illustration of the ways in which the concept of 'community' is mobilized to govern and control, surely this is it.

Finding 'graffiti solutions' in Sydney

If most would agree that graffiti has become a part of the urban fabric in many cities, not all would celebrate this fact (see Figure 4.2). Certainly, when the word graffiti appears in policy and planning documents, it rarely features as an aspect of the public domain that enhances our 'quality of life'. The appearance (and re-appearance) of graffiti (along with fly posters, stickers and other unauthorized modifications to urban surfaces) is a constant frustration for regulatory authorities and property owners. Many users of public space report feelings of insecurity when confronted by graffiti. According to advocates of contemporary agendas for improvements to urban 'quality of life' and the regeneration of urban public realms, graffiti contributes to the degeneration of neighbourhood, commercial and civic spaces. Following the 'broken windows' theory of crime prevention, these advocates argue that apparently small acts of vandalism such as graffiti send a signal that the community does not care about its local environment. If left unchecked, so the story goes, graffiti can therefore result in a downward spiral of neglect as more serious forms of crime and anti-social behaviour take root (Kelling and Coles 1997).

This framing of the 'graffiti problem', it should be noted, has a history. In New South Wales, a 1986 report by criminologists Wilson and Healy for the State Rail Authority argued that there was in fact no statistically observable connection between graffiti and personal safety – rather, such connections were

4.2 Graffiti tags and stencil, Sydney.

Source: Kurt Iveson.

a matter of perception and thereby open to public intervention. The report noted that:

> [B]oth staff and public must accept a certain level of rail vandalism and graffiti as inevitable – which they are. However, such acceptance is problematic – and probably unlikely – as long as vandalism and graffiti are seen as closely associated with violence (or even indicative of threatened or potential violence) on State Rail.
>
> (Wilson and Healy 1986: 64)

As a consequence, they urged both State Rail and political leaders to make the distinction between graffiti and violence 'clearly, frequently and publicly' (Wilson and Healy 1986: 64). Nonetheless, in New South Wales as elsewhere, political rhetoric about law and order works to precisely the opposite effect, continually asserting the connection between graffiti and community disorder in order to justify a range of ever-harsher regulatory responses.[3]

A range of solutions to the 'graffiti problem' have been explored in cities around the world. In Sydney, as in other places, we have seen attempts to establish circuits of exclusion to deal with the threat of anti-social graffiti writers. Vast resources are spent on the 'war on graffiti'. By 2000, State Rail estimated that it spent A$3.5 million each year in an effort to ensure that all graffiti on trains and inner metropolitan train stations were removed within 24 hours, while the Rail Access Corporation, which controls rail corridors, spends around A$2 million annually on graffiti removal (Geddes 2000: 12). Known graffiti targets are hardened through the development and application of graffiti-proof materials, surveillance and spatial barriers such as razor-wire fencing. Legislation is introduced to criminalize the sale of aerosol paint to minors, and harsher penalties are introduced as both deterrent and punishment for those caught writing graffiti. In New South Wales, even the possession of a spray can with intent to commit a crime is an offence after the passing of Summary Offences and Other Legislation (Graffiti) Amendment Bill 1994. Specialist police units such as the NSW Graffiti Task Force (now disbanded) have been set up to gather intelligence on graffiti writers and their activities. Local councils and other urban authorities such as the Rail Access Corporation have been given special powers to remove graffiti from private property where property owners have not taken action themselves. These measures are typically lauded by observers in the mass media, who are only too happy to support the government's war effort.

Of course, the war has not been won. Graffiti continues to appear on urban surfaces in Sydney and in other places where exclusionary measures have been applied to the 'graffiti problem'. The spiralling costs of the war have made room for some to propose other solutions to the problem. A more inclusionary approach has also taken shape involving the provision of opportunities for 'legal graffiti' or 'aerosol art'. In New South Wales, under the rubric of the State

Government's 'Graffiti Solutions' policy, a range of community organizations have successfully applied for funding under the 'Beat Graffiti' grants scheme to organize local projects designed to provide graffiti writers with alternative legal avenues for their work. The legal graffiti projects funded under the Beat Graffiti scheme have taken a range of forms – some involving the basic provision of legal spaces for graffiti, others involving more sustained attempts to employ established graffiti writers to run workshops with aspiring younger graffiti writers to develop their artistic and entrepreneurial skills. Advocates of legal graffiti typically emphasized the cultural dimension of hip hop style graffiti, in an attempt to de-couple the association of all graffiti with criminal and anti-social intentions. This notion has become more widely accepted as graffiti-style design and artwork becomes more and more ubiquitous in corporate advertising and products intended for the 'youth market'.

In seeking to secure a place for graffiti in the city through government-funded programmes, advocates of this more 'inclusionary' approach have been forced to accept the dominant framing of (illegal) graffiti as a 'problem' for urban policy. Their proposals for funding, and the 'success' or otherwise of their programmes, are measured against a yardstick established in advance by the State Government – the ultimate eradication of illegal graffiti. That is, 'legal walls' are valued to the extent that they can be proven to 'beat graffiti' in other parts of the locality in question. Some legal graffiti programmes are very self-consciously aligned with wider efforts to eradicate illegal graffiti. Participation in some programmes is premised on a renunciation of all other forms of graffiti, and in some cases on the provision of intelligence to programme coordinators about the activities of 'illegal' graffiti writers, which is then passed on to police in order to assist in arrests. As Andrew Collins, a youth worker at a Police Citizens Youth Club in Newcastle, put it, 'strategies run through arts bodies or youth organizations provide an ideal intelligence source for law enforcement bodies' (Collins 1997). Here we have a classic case of social control through inclusion.

Of course, this is not to say that there have been no other outcomes from these funded programmes. Other legal graffiti programmes have taken a less exploitative approach, remaining resolute in their attempts to secure the trust of young people by keeping their programmes off limits for law enforcement bodies. In the process, they have pointed towards the potential contribution legal graffiti programmes might make to *engaging* young people in a wider politics of urban aesthetics in which the graffiti 'problem' is redefined. With their emphasis on skills development and the circulation of knowledge about hip hop culture and graffiti styles, the more progressive legal graffiti programmes may enable a counter-discourse about graffiti to emerge, in which *bad* graffiti is the problem to be addressed, not graffiti per se. Aspiring graffiti writers are given opportunities to hone their skills, to debate graffiti aesthetics and ethics, to develop their craft and to pass on their knowledge to other young people.

In this way, rather than simply being enlisted to 'solve' the graffiti 'problem', these young people may yet begin to take part in debates about what the city should be like, and what it could be like.

Conclusion

The challenge of finding alternatives to the 'exclusion' of young people from city spaces is not best met with urban policy prescriptions premised on 'inclusion'. Indeed, I have argued in this chapter that efforts to enmesh young people in circuits of exclusion *and* inclusion are characterized by distinctly undemocratic understandings of 'the city' and city life that are characteristic features of neo-liberal forms of urban governance. Here, the values of 'the community' and the dimensions of the 'good city' are considered to be written in stone, agreed upon by everyone in society with the exception of the 'anti-social'. The 'anti-social' are therefore to be worked upon, brought into line with community values. Urban policy debates are too often restricted to finding the best 'solutions' to 'problems' that are agreed upon in advance – the carrots of inclusion or the sticks of exclusion? As Stanley Cohen noted some two decades ago, at a time when the gradual shift to new forms of neo-liberal urban governance was beginning to take shape, inclusion and exclusion often work hand in hand in efforts to achieve social control:

> At their purest, these forms of inclusion work because they are voluntary or simply because they are not recognized to be social control. But they require a back-up sanction: if you do not take the initiative yourself or if we do not spot you in advance, this is what might happen to you.
>
> (Cohen 1985: 233)

And while we might prefer the carrot to the stick, Cohen warns us to interrogate this preference more critically:

> It is by no means obvious that exclusion must be the more intensive and less tolerant mode. We might separate a group only to ignore it completely, while inclusion might entail massive efforts to achieve normative or psychic change.
>
> (Cohen 1985: 219)

My critique of exclusion and inclusion should not be read as an attack on those community workers who advocate more 'inclusionary' policy responses to urban problems such as 'anti-social behaviour'. Indeed, my argument in favour of *engagement* with children and young people as an alternative to social control is informed by many conversations with such workers and young people – most of who are all too conscious of the 'citizenship games' they must play in order to take any part in urban politics as it is currently framed. Just because they play the game does not mean that they have to like it. Their own critiques of the game, and their efforts to occasionally step outside it, point towards a politics of engagement with young people that, I believe, constitutes an alternative to social control as the goal of urban policy and politics.

Engagement, as I have developed the concept in this chapter, is distinguished from social control by its different understanding of the city. Efforts to engage young people – such as *GiBBER* and some legal graffiti programmes – are either explicitly or implicitly premised on the notion that the characteristics of the 'good city' are not settled in advance of politics and policy. Engagement is premised on the notion that debates about the characteristics of the good city should be the very stuff of urban politics and policy. As other contributors to this collection have also argued, children and young people must have genuine opportunities to participate in these policy debates. Constructing a politics of engagement with young people is therefore a matter of building common projects that unsettle dominant notions of the good city or community values. Rather than requiring young people to put aside their anger and frustration in order to be 'included' in the city, a politics of engagement might even be energized by youthful interventions in urban politics that refuse to accept those aspects of urban life that are too often taken for granted as the natural order of things.

Notes

1 In Australia, see for example the Children (Protection and Parental Responsibility) Act NSW 1997, the Western Australia Parental Support and Responsibility Bill 2004. In the UK, parenting orders and parenting contracts are legislated for in the Crime and Disorder Act 1998 and the Anti-social Behaviour Act 2003.

2 These two cases are drawn from the more detailed research that forms the basis for two chapters in my forthcoming book, *Publics and the City*.

3 Joe Austin (2001) makes a similar point about the way in which graffiti writers were initially treated as popular folk heroes in the New York press when graffiti first took off in the early 1970s, before the consolidation of the graffiti-writer-as-antisocial-criminal rhetoric later in the decade.

References

Austin, J. (2001) *Taking the Train: How Graffiti Became an Urban Crisis in New York City*, New York: Columbia University Press.

Australian Labor Party (NSW) (1994) *Law and Order Policy*, Sydney: Australian Labor Party (NSW).

Braithwaite, J. (1989) *Crime, Shame and Reintegration*, Sydney: Cambridge University Press.

Carr, B. (2002) Speech by the Premier Bob Carr at the Law Society, 24 October. Available: www.communitybuilders.nsw.gov.au/download/premspeech.pdf (accessed 2 May 2006).

City of Perth (1995) *Perth – A City for People: Strategic Management Plan, 1995–2000*, Perth: City of Perth.

City of Perth (1997) *Youth Forum: Report of Findings*, Perth: City of Perth.

Cohen, S. (1985) *Visions of Social Control: Crime, Punishment, and Classification*, Oxford: Blackwell.

Collins, A. M. (1997) 'Hip Hop Graffiti Culture (HHGC): addressing social and cultural aspects', in *Graffiti: Off the Wall seminar proceedings*, Sydney: University of New South Wales, 11 June. Available online: www.graffiti.nsw.gov.au (accessed 25 August 2005).

Collins, D. and Kearns, R. (2001) 'Under curfew and under siege? Legal geographies of young people', *Geoforum*, 32 (3): 389–403.

Crawford, H. (1997) '$3m bid to keep louts off trains', *The West Australian*, 7: 21 July.

Dean, M. (2002) 'Liberal government and authoritarianism', *Economy and Society*, 31 (1): 37–61.

Gallop, G. (2003) 'Premier unveils Northbridge curfew policy', *Press Release*, 26 June.

Geddes, W. (2000) 'Battle to beat graffiti', *Daily Telegraph*, letter to the editor, 20 January: 12.

GiBBER (1995) No title, in *GiBBER* Magazine, 7: 1.

Government of Western Australia (2003) *Government Achievements Report*, 24 March–28 April, Perth, Government of Western Australia.

Jefferys, E. (no date) 'Censorship', in *Proceedings of DARE Conference*. Available online: www.orca.on.net/dare/jeffreys.html (accessed 25 August 2005).

Kelling, G. and Coles, C. (1997) *Fixing Broken Windows: Restoring Order and Reducing Crime in Our Communities*, New York: Simon & Schuster.

McNamara, G. (1994a) 'Police push parents' rights', *The West Australian*, 26 May: 9.

McNamara, G. (1994b) 'Business to act on crime', *The West Australian*, 25 May: 45.

Manton, C. and Magil, P. (2003) 'Gallop blasts curfew critics: aboriginal leaders accused of leading their children down the wrong path', *The West Australian*, 21 October: 3.

Office of Crime Prevention (2004) *Responsible Parenting Orders and Responsible Parenting Contracts*, Perth: Department of the Premier and Cabinet, Government of Western Australia.

Peck, J. and A. Tickell (2002) 'Neoliberalizing space', *Antipode*, 34 (3): 380–404.

Prior, N. (1997) 'Dancing to a united beat', *The West Australian*, 21 March: 11.

Rose, N. (2000a) 'Governing cities, governing citizens', in E. Isin (ed.) *Democracy, Citizenship and the Global City*, London and New York: Routledge: 95–109.

Rose, N. (2000b) 'Government and control', *British Journal of Criminology*, 40 (2): 321–39.

Sercombe, H. (1995) 'The face of the criminal is Aboriginal: representations of Aboriginal young people in the West Australian newspaper', *Journal of Australian Studies*, 43: 76–94.

Valentine, G. (1997) '"Oh yes I can." "Oh no you can't": children and parents' understandings of kids' competence to negotiate public space safely', *Antipode*, 29 (1): 65–89.

Wilson, P. and Healy, P. (1986) *Vandalism and Graffiti on State Rail*, Canberra: Australian Institute of Criminology.

Part Two
Policies, professionals and the environment

Chapter 5

Colliding worlds

Planning with children and young people for better cities

Claire Freeman

Children and the planning connection

Planners play an important role in shaping the environment. They take responsibility for guiding and managing the development of cities, and for shaping the future directions of city development. Planners focus on land use and work primarily in development control, policy, strategic and forward planning, and can be found working at national and local government level, in private consultancies and in governmental and non-governmental agencies. Planning and planners influence children's lives and experiences. The policies they implement, the design ideas they support and the decisions that they take all manipulate and shape the environment in which children live, and thus materially impact on the quality of children's lives. Planners outside specific child or youth oriented council initiatives seldom take children's needs and views into account in development or policy planning. They are neither trained to work on behalf of children nor with children, though children are possibly even more influenced by the nature of their environment than their adult counterparts. Yet the decisions they take, the policies they implement, and the developmental trends they support do have direct impacts on children's lives and well-being. Planners need to be aware of children and young people in the way they plan the environment and to be aware of how the decisions they make impact on children. In order to do so, planners need to have a better understanding of children's lives and societal relationships and to have access to and support from child and youth centred policies and practice. In the discussion of physical environments, the emphasis is on environments that work for younger children. Where discussion relates to older children, the term 'young people' is used. This chapter then explores the relationship between planning, social and environmental change and children and young people's lives. It looks at how changes and developments in cities over time have impacted on their lives and offers positive ideas on how future planning can be improved for children and young people.

The chapter is called 'colliding worlds' as too often this is what happens. Planners act on behalf of the public and plan for the 'public good'. However, in practice what planners believe are in the interests of the public, as is the case with recent transport, zoning and other planning ideals, in fact often inhibit and frustrate children's and young people's lives. Planners have become aware of the need to plan for and with children and young people. 'Collisions' can indeed be avoided by physical separation, but they can also be avoided by coming to understand divergent views, different needs and developing ways forward that recognize and prioritize the needs of all groups in society, especially those whose voice is only now beginning to be heard. The discussion and examples focus on the Western city, particularly those in the United Kingdom and New Zealand, two countries where the author has worked in the field of children and planning. The chapter begins by exploring the changing physical nature of the urban environment and the impacts of these changes on the lives of children and young people. The second part examines the challenges and possible ways forward in developing better planning policy and practice for children and young people.

Why do children and young people matter?

Public participation is at the heart of good planning process, and is universally acknowledged as a 'good thing' for planners. But only in the 1990s would children and young people be added to the list of groups to which planners should give special consideration in the planning process. The primary catalyst for this belated inclusion was the UN CRC (UNICEF 1992). Three of the convention's 54 articles are particularly important, as they provide the basis on which the other rights are built. These are:

Article 2 establishes that all rights identified must be available to all children and young people, without discrimination of any kind;

Article 3 requires that in all actions concerning a child, the child's 'best interests' should be a primary consideration;

Article 12 states that in all matters affecting them, children and young people have a right to express their views.

These rights are thus far reaching in scope, covering all areas of children's lives, and they have particular relevance for those responsible for the environments in which children live their lives. Implicit in the convention is a clear understanding that children's environmental, social and participation rights will be acknowledged and supported. There has also been a more general and growing recognition of children's rights in society. It is a move that has found much support in both central and local government. In New Zealand, which initially was slow to embrace this growing agenda, central government is now paying increasing attention to children and young people. This can be seen in the recent production of a number of strategies and guidelines directed at children and young people: *New Zealand's Agenda for Children* (Ministry of Social Development 2002), *Youth Development Strategy Aotearoa* (Ministry of Youth Affairs 2002) and a number of associated publications such as *Making it Happen: Implementing New Zealand's Agenda for Children* (Institute of Public Policy 2002) and three central government participation guides directed at those working with children and young people (Ministry of Social Development 2003, 2004; Ministry of Youth Affairs 2003). The intention is for these central government documents to be taken up by the voluntary sector, the private sector and, most importantly, the public sector, where they will be adopted across sectors and by a range of agencies. Planners are key actors in local government, and for them there are a number of overpowering arguments as to why they should embrace the promotion of children's rights and engage with children and young people. These include the fact that children and young people have the right to be included, a right that has legal and moral underpinnings. The community benefits from their involvement as partners in development. Children and young people offer different perspectives and have different needs. Bringing them together with other adults, and with professionals, facilitates understanding of their differences. It also allows for the exchange of views and an understanding of what constitutes a good environment. The broader the participation, the greater the benefits for everyone.

What makes a good environment?

One, if not *the*, prime motivation for planning is the creation of good environments. This begs the question of what is a good environment. The following characteristics were identified by Driskell (2002: 26), from the findings of the UNESCO-MOST 'Growing up in Cities project', as contributing to the success and failure of cities from a children's perspective:

Positive indicators	*Negative indicators*
Social integration	Stigma and social exclusion
Variety of interesting settings	Boredom
Safety and freedom of movement	Fear of harassment and crime
Peer meeting places	Racial or ethnic tension
Cohesive community identity	Heavy traffic
Green areas	Uncollected rubbish and litter
	Lack of basic services
	Sense of powerlessness

In thinking about and designing better environments, the ideal as currently practised by planners is not necessarily better for all. The ideal when conceived by adults is 'better' mainly for adults, though children may by default also be beneficiaries. It is said that a city that works for children and young people works better for everyone. Well, almost. The elderly and skateboarders are unlikely to reach a shared consensus on public space. Environmental design frequently involves making choices between the needs and values of different groups. These choices are especially pronounced in open or public space design, where the needs and values of different age user groups are frequently in direct conflict (Francis 1988: 67). Related to conflict has been the process of exclusion, particularly around the access to, and use of, public space by children and young people. As White states '[a] good living environment for the majority who make up the community is not one which is based on principles and practices of social exclusion' (White 1996: 37). What then, are some of the processes that have occurred and that are still occurring that have contributed to this developing exclusion of children and young people, and how has it been possible within the context of an ongoing planning commitment to building better environments?

Changing nature of society and the context of childhood/ environment relations

Physical change

While not attempting to idealize urban history, which has never been kind to children, the potential for such conflict has clearly increased and been exacerbated by some of the planning and design principles followed in the latter part of the twentieth century. The nadir of planning with regard to children came in the 1950s and 1960s, with the wholesale redevelopment of neighbourhoods into high-rise block housing estates. These were, and continue to be, environments that no self-respecting planner or architect could ever argue had children's interests at heart. Their legacy continues to fragment and damage the lives of children in many cities around the world today. One significant example was the move in Europe in the 1960s to replace the decaying, cramped terrace and tenement housing, often referred to as 'slums', with something more 'healthy'. The move did take children out of streets and neighbourhoods characterized by poor physical conditions, but what was overlooked was the fact that these neighbourhoods often also had a rich social life (Young and Wilmot 1979). The move was frequently into the suburbs and urban high-rise estates, characterized by social fragmentation. In many cases the estates soon degenerated into equally deplorable, and often unhealthy, social and physical environments.

The redesign of the Western city over the last century – in particular after the Second World War – and the development of new ways of using the city have impacted significantly on the lives of children. Transport has been one of the key influences. The development of transport systems and private vehicles has led to a separation of the connection between home and places of work, schooling, leisure, recreation and shopping. For many, the neighbourhood is no longer the place where these activities occur but may, as in the case of work, only be accessed following a commute by private vehicle that can take over an hour. In many larger cities, congestion means that the commuting parents may attempt to avoid traffic chaos by timing their morning journey to start earlier and delay their home journey until later, with the result that the commuter is present in the home for little of the child's waking day.

Private cars have given parents greater flexibility in choosing schools. Many parents choose not to send their children to the local school but to a 'better' school, or a more convenient school (e.g. one close to work or that has after-school childcare). Thus, not only schooling but also visiting of friends and taking part in school sports outside of school hours become increasingly dependent on the car and further removed from the child's own neighbourhood. The development of the suburb, the out-of-town shopping centre (so characteristic of Australia), the business park, the leisure complex and the medical centre have all acted to entrench this locational divorce between where children live and where the activities that they and their family take part in are geographically located. Children's lives become a fragmented mosaic of places – school, child-care, club, shops and playground lacking the linkages provided by walking along streets, past familiar people and places. As Lennard and Crowhurst-Lennard state '[i]t is difficult to develop a sense of meaning in a human or physical landscape that does not make any sense' (1992: 38). Older towns, cities and neighbourhoods in both Europe and Australasia that have avoided the planner's and developer's attention often still retain the mix of activities within neighbourhoods. This has traditionally provided opportunities for children in their daily lives to experience and independently access the breadth of urban life. The geographical separation of activities also reduces the opportunity for children to develop their independence. Their reliance on adults (usually parents for lifts) means that children remain in a state of adult dependency for much longer than occurs when they can walk to friends, clubs or shops, and their social lives are grounded in the local neighbourhood. It can be argued that the redesign of our cities in recent years, the planner's fixation with zoning and the prioritization of cars over people has created divisive city environments in which the child has become marginalized and excluded.

Current trends – such as the support for private transport and related traffic 'control', urban intensification, which reduces open space and concertinas more cars into a smaller area, and the development of gated communities – also impact on children's health, mobility and social development. Now planners are being forced to re-evaluate what constitutes good urban form, and much positive thinking is evident in the new urbanism ideas, and in regeneration initiatives such as the South Bank Parklands redevelopment in central Brisbane. South Bank is a child and family friendly extensive urban development, focused around free and safe play with plenty of stimulating micro-spaces. It is also a real and active public realm that hosts child- and youth-focused activities. The playgrounds, the beach and waterways and large open spaces are all conducive to skateboards and roller-blading. It also has good multi-modal public transport access, especially important for older children travelling independently. The South Bank example is exceptional rather than typical, and children and young people are not usually a significant consideration when new directions in planning and new planning projects are embarked on. The physical environment does impact on children's lives, and good urban form can act as a catalyst for reintegrating children into the local physical and social environment.

Excluding environments

The physical environment impacts on children's lives, but so too does the social environment. Beliefs about what is good for children, beliefs about the interaction between children, family and wider society, beliefs about what constitutes 'good parenting', and beliefs about the role children play in society are all part of their social environment. Children's removal from the outdoor environment, with its attendant loss of independence, has been the subject of a growing body of research (Tranter and Whitelegg 1994; Freeman 1995; Cunningham, Jones and Barlow 1996; Valentine 1996a; 1996b; Tranter and Pawson 2001; Kearns and Collins 2003). A recent study of children's changing environmental access is that by Pooley, which in many ways provides a welcome, if perplexing contrast, in its findings. Pooley, Turnball and Adams examined changes in patterns of everyday mobility across three generations (from the 1940s) living in cities in northern England. The study found that 60 per cent of trips taken by 10- and 11-year-olds were still undertaken by foot. They did, however, find the predicted overall decline in walking and bus travel, a decline in trips unaccompanied by adults and for more restricted distances, noting that 'today few children are allowed to experience and negotiate everyday risks: parents curtail their playspace in a way that seemed not to occur even a decade ago' (Pooley, Turnball and Adams 2004: 8). For many, especially younger, children their removal from the public environment is most immediately noticeable in their removal from the area around the home and the street. The street is particularly important as a space in which children and young people come together: it constitutes an important cultural space for young people, and a space that they imbue with their own meaning. Matthews' (2002) study of the street as 'thirdspace' found that children and young people rely heavily on it as playspace, a place for meeting, talking, leisure, sport and shopping, and a space that provides security, freedom and social opportunity. Any loss of access to this 'thirdspace' is one that has serious repercussions for children and young people.

Why is it more difficult for children and young people to access the public spaces in their neighbourhood? The first reason has already been discussed. The redesign of our cities with its separation of activities forces reliance on transport as distances between activities become too great for easy walking, and enhance parental anxiety about unaccompanied children. The second reason of recurrent concern to parents is the very real danger posed by cars to children's safety. Valentine (1996b), in her study of the changing spatial restrictions of childhood in the UK, found 34 per cent of parents of children aged 8–11 rated cars as the most significant danger for their children. The response to the very real danger posed by traffic has been interesting. Researchers, planners, traffic engineers and others with responsibility for managing our urban centres 'have tended to conclude not that motor vehicles should be strictly controlled in cities, not that driver education should be a first priority, but that children should be constrained, kept indoors, accompanied or policed by parents' (Davis and Jones 1996: 233). It is an approach that confirms the belief that children, not cars, are the problem. The 'What Works for Children' organization, in its traffic calming poster, provides a number of traffic related facts, including the following:

- Children in poor neighbourhoods are five times more likely to be injured by a car than those in affluent areas (Wazana *et al.* 1997).
- Children whose families have fewer resources tend to live in more dangerous housing and road environments, have fewer safe places to play, and go out on foot more often (Roberts and Power 1996; Roberts, Norton and Taua 1996).
- When a child is hit by a car travelling at 40 miles per hour, only 15 out of 100 will survive; when the car is travelling at 20 miles per hour, 95 out of 100 will survive (Child Accident Prevention Trust undated).

What is interesting in these three examples is that the child's well-being is consequent upon adults: their economic resources as in the first two, and the speed of their driving in the third. If society had children's well-being and safety as a priority in its cities, all three could be addressed to their benefit, but to do so would require significant challenging of the predominant ways in which society works. It would require all children to have good living standards in their homes and neighbourhoods, and the reduction of the right for adults to travel at speeds that can harm children.

The third reason for removing children from the street environment is the increasing feeling of fear related to children in the public environment. Stranger danger and the fear of child abduction were rated by 45 per cent of the parents in Valentine's study as the greatest danger they feared for their children. There has been the rise of a strong and pervasive view of children as vulnerable, in part caused by a small number of abductions and other incidents that have been given a high media profile, especially in the 1980s. The growth of the image of the child as 'vulnerable' has contributed markedly to the view that good parenting is about keeping children safe and knowing where they are at all times. As a consequence, childhood is becoming increasingly controlled (Hill and Tisdall 1997).

Not only has the public environment been seen as threatening for children, but conversely children, primarily older children or 'teenagers', are themselves seen as a threat. As Boyden reports 'The moral panic about the urban young is not confined simply to their apparent criminality, but also to what is feared to be their innate subversiveness' (Boyden 1991: 37). For many adults, the mere presence of teenagers on the street is seen as threatening, and as Woolley *et al.* (1999) found, threatening for some children too. 'Children . . . give vivid accounts of their perceived threats, from some adults on the street, from older adolescents, and from groups out of town' (Woolley *et al.* 1999: 287). Has this always been the case?

For a long time there has been concern about young people hanging about. 'Long haired layabouts, teds, mods and rockers, yobbos' and other terms were used by older generations to refer to what they saw as dissolute young people 'hanging about on street corners'. Why do young people 'hang about'? Obviously by doing so they can easily meet up with friends and see what is happening in the world. Unlike other groups in society, young people have no public places that they can legitimately access and call their own. The way that young people use space can put them at odds with other users. As Valentine notes '[y]outh in their teens take possession of the city and use its public spaces in a more free and intense way than do many other groups' (Valentine 1996b). Young people have become increasingly ostracized from public space and subjected to increased policing as space has become increasingly commodified. In spaces such as the shopping mall and the leisure complex, young people can be made to feel especially unwelcome (White 1996; 2001; Hil and Bessant 1999). Chapter 4 by Kurt Iveson identifies a number of actors and agencies engaging in the social exclusion and control of young people, together with the construction of what he terms 'circuits of exclusion'. Public space is vitally important to children and young people as social space, yet their right to use this space is subject to significant and growing social and physical restraints (see Figure 5.1).

Does exclusion matter?

The answer is an unequivocal yes. It matters because the public realm is where children and young people learn about society, where they explore it, observe it, absorb its values, and gain a sense of belonging. Life cannot be learnt within the confines of the family, the school, playground or youth club. A policy of setting aside child spaces (the current dominant planning policy for children) can meet some physical needs, but not wider social and developmental goals. For young children, as Wheway and Millward found out: 'The quality of the

5.1 Capturing public space, BMX riding in a city square in Brisbane.
Source: Claire Freeman.

environment within two streets of the front door is extremely important' (1997: 33). An 'important factor for children was the desire to be part of the community being where it's at', a fact that explains why, in their study, it was found that back gardens are hardly used at all – yet front gardens are very popular (Wheway and Millward 1997: 33). This finding identifies the fact that children want to learn how they, their home, their street and their neighbourhood fit into wider society. Only if they have access to this wider society as independent participants can they learn how their community works and their place within it, what is safe and what is risky, who to trust and not trust, what is usual and unusual, how to anticipate events and how to deal with the new and the unfamiliar as they occur. As Crane and Dee put it, a 'viable community' is not an inclusive one that values diversity but one where people 'actively gossip about each other, know one another and spend their time on front steps watching passers by' (Crane and Dee 2001: 12). Research undertaken in small towns in New Zealand supports this view, as young people do not report the same level of alienation and distrust from adults as has been noted in cities. This can probably be accounted for by the fact that in small settlements, the young people and the adults know each other (Williams 2000; Panelli, Nairn and McCormack 2002).

Removal from the streets and neighbourhood removes children and young people from the experiences such environments offer. It results in a loss of those skills and benefits associated with independent mobility, the ability to develop autonomy and independent decision-making. These skills include making judgments about what is safe, time management, the development of social responsibility and caring for others, especially the young and more vulnerable in the group. Their removal also impacts on society as it removes from adults a sense of responsibility for children and young people, the loss of the community role in looking out for their safety and for socializing with them in the community. The impacts of this are being addressed in some places. Legally introduced in the Netherlands in 1976, the 'Woonerf' (home zone) concept is one that is finding much support, particularly in Austria, Denmark, Germany and latterly in the UK and Australia (see Engwicht's work on street reclaiming). Home zones are residential streets that have experienced traffic calming so that they are safe for children to walk, cross or play on. The success of the venture has been such that 136 home zones are listed on the Five Roads Forum website on home zones in the UK (Residents' Association for Five Roads Home Zone 1995). Perhaps it is time we reassessed the way that neighbourhoods and streets in Australasia work, and reasserted their role as community meeting spaces, taking up Engwicht's challenge for people to 'Reclaim the Street' from the perfidious motor car (Engwicht 1999).

For planners the challenge of creating child friendly cities is waiting to be taken up. While there is much debate and some uncertainty about planners' roles with regard to the social construction of society and with social well-being, what is clear is that planners – through policy, practice and their decisions – do influence social well-being. The final section of this chapter moves away from discussion of the physicality of child friendly cities to look at how processes can be developed that make it possible for the interests of children and young people to be considered and represented in environmental decision and policy making.

Developing policy and practice that benefits children

To date, children and young people have not been seen as a priority, or even a significant consideration, by planners. They have been accorded low political status and been largely ignored in the public policy environment. If decision makers are to redress this position of marginality, they will need to acknowledge children's interests as valid and be prepared to represent them and make provision for children's autonomy. The intention in addressing children's interests is not to offload political responsibility on to children but to 'create a society in which, consistent with fundamental legal values, there is respect for children as persons' (Melton 2000: 144). Up to now there has been a tendency to see children's rights as exclusionary, precluding or taking over the rights of parents, teachers or other adults. Such an approach fails to recognize that children are integrated into society and that decisions that affect children also affect others. While age, competency and relationships with others are important in addressing children's rights and autonomy, a key issue for policy makers is how policy and practice can be designed so that children are a key consideration. The process will vary depending on societal context, age, competency and other factors. Indeed a process of 'graduated' decision-making may be the way forward, with children and adults working and learning together. In Australia, New Zealand, the United Kingdom and elsewhere, processes and procedures are being introduced at local government and school level to provide increased opportunities for children and young people to develop shared decision-making. As Melton asserts: 'the strategy of intergenerational decision making can serve as the foundation for a new communitarian ethic in the various settings of which children are a part' (Melton 2000: 155). Such shared decision-making can represent an important first step on the path towards achieving Melton's communitarian ideal. To achieve this ideal, developments in the more formal settings of school and local government need to be linked to similar developments at the micro level, i.e. at the level of decision-making as it pertains to children's everyday lives.

The micro level has been the focus of much research regarding issues such as children's independent mobility and play ranges, as already discussed. The importance of 'everyday spaces in and through which children's identities and lives are made and remade' is becoming recognized as essential to developing and understanding children's social and environmental well-being (Holloway and Valentine 2000: 11). However, less well known is the part children play in shaping their own experiences, particularly the role of decision-making in their everyday lives. This can be seen in the example of 'walking school buses', a rapidly growing phenomenon in New Zealand cities. Kearns and Collins in Chapter 7 explore their development in Auckland, New Zealand. We know that there has been a phenomenal growth in numbers in recent years, but little is known about children's autonomy as it relates to the decision to become part of the walking school bus. It is important for decision makers to recognize the possible disjuncture between participation levels in different arenas of children's lives. Shared decision-making in the local government arena is unlikely to be successful if children have no experience or expectation of shared decision-making in other aspects of their lives, and vice versa. Children, given the opportunity, are extremely versatile and skilful negotiators of different contexts, and are adept at transferring skills learnt in one context into another context. The aim should be to enhance children's autonomy in all aspects of their lives.

In addressing the growing agenda pertaining to children, the remit of planning policy must also be addressed in its widest sense. Children do not fall into a neat planning enclave; their interests traverse sectoral, professional, government departmental and agency boundaries. For planners, children are usually considered within the confines of their educational and recreational needs. But children permeate the whole environment, for, as Hil and Bessant state:

> The whole city is a play space, not just the small specific plots of land set aside for children's use. In their exploration children are entitled to freedom from traffic dangers, from danger to their person from other people and from overbearing adult control of their play activities . . . A sense of community must be a planning goal.
>
> (Hil and Bessant 1999: 129)

Planners need to understand children's wider use of the environment (see Figure 5.2). Transport, housing and shopping are all key arenas of children's lives needing special consideration, as are the interactions between them, for example, the link between public transport and the location of leisure centres.

Developing appropriate frameworks

Once the groundwork in terms of acceptance of the principle that children should be considered in all aspects of planning and decision-making has been achieved, the next stage is to develop appropriate institutional frameworks. One of the problems to date has been that local and central government, together with a range of interested agencies, have launched into child and youth projects and initiatives with limited understanding. They have set up forums and councils, developed strategies and policy documents, and designed participative initiatives, without adequate consideration of what inclusive practice really means, and without establishing the necessary groundwork. Thus, many well-intentioned projects have foundered or not had influence beyond the lifespan or the boundaries of the project. In our research on planning with children and young people in New Zealand, we looked at what was happening in planning in local government (Freeman, Aitken-Rose and Johnston 2004). Although many councils had set up skate parks and involved young people in the designs, this was not connected to any consideration of how they might be involved in developments relating to other parts of their lives such as transport, housing, shopping or leisure developments. Compartmentalization occurs at many levels. Our research also found a lack of connectivity between central government participation initiatives, local government activities and action by planning practitioners. Planners were often unaware of what was happening at central and local government level. Where councils had child and youth policy documents and/or advocates, few planners made any use of these resources. There is then a strong need to develop networks and a better exchange of information between different levels of government, between agencies and between planners. The current practice of ad hoc, isolated initiatives has limited effectiveness and little impact outside the initiative itself. Initiatives need to be developed with respect to the different,

5.2 An inventive three-year-old exploring the environment, climbing railings along a city street in Dunedin, New Zealand.

Source: Claire Freeman.

but related parts of children's lives, and to building long-term change – outcomes not possible through isolated one-off events.

When asked about processes and the potential for input from children and young people, many planners indicate that they are well aware of the need to improve current practices. They are conscious of the necessity of adopting a different approach and the limitations of existing adult oriented methodologies. Planners, despite their wishing to engage with children and young people in more appropriate ways, are uncertain about how to achieve this, finding themselves ill equipped in terms of their training and resources. If progress is to be made, planners need to engage with the wider debates on children, young people and society, and to embrace the opportunities for more participative and visionary planning. To achieve these goals is a major challenge for planners, as they will have to enter into areas for which their training and expertise has not prepared them. To do justice to the developing children's agenda, planners will need to:

- be able to understand how both current and future planning developments, designs and processes impact on children;
- recognize the changing and complex nature of childhood;
- develop an understanding of children and young people's environmental experiences;
- understand what it is that children and young people want; and
- be prepared to work with and on behalf of children and young people.

There is one major resource that can assist planners willing to take up this challenge and realize the opportunities: children and young people themselves. Research with children and young people on their experiences indicates quite clearly that they do have understanding, views and ideas relevant to planning and to the improvement of planning practice. It has been suggested that a problem area is that young people in particular do not want to work with adults. Our research (Freeman, Nairn and Sligo 2003) undertaken with young people does not support this view. We found young people do want to work with adults and recognize the benefits of doing so, but they want to be treated as respected partners, to have their views, skills and ideas respected. The issue is not that young people do not want to work with adults but that their experiences of working with adults have not always been positive ones (see Morrow 1999; Smith *et al.* 2000; Smith *et al.* 2002; Freeman, Nairn and Sligo 2003; Nairn, Panelli and McCormack 2003). Unless planners have a really good understanding of children and young people, and are prepared to listen to them and to change planning processes to meet their needs, then these negative characteristics will be perpetuated.

A new approach

Planners have experience of working with the community but only recently have any planners started working with and on behalf of children and young people. The arguments for developing cooperative planning approaches between children, young people and planners have become hard to ignore. Pressure to give greater recognition to children and young people has come from actions at governmental and intergovernmental level. It has also come through the signing of treaties and conventions, the development of strategies and policy documents, the appointment of officers with specific child and young people remits, children's commissioners and child and youth advocates, the development of youth forums and councils, and from other initiatives. Strong messages are also coming from both practice and research asserting the benefits of child and youth centred approaches to planning policy making and development.

For society, the benefits are in encouraging participation across the age groups, valuing contributions and building a positive, democratic society (Ministry of Social Development 2003). For children, the benefits relate to their own developing confidence, skills, knowledge, understanding, the development of relationships with adults and the development of facilities, services and decisions that are child centred and aware of their needs. For planners and other professionals, such inclusion means a shift in the focal point of planning and decision-making towards people at the local level who are most affected by the decisions made. While such a shift can be challenging for planners, especially when the key players are children, the benefits associated with the shift should be sufficient for planners to overcome any qualms they may have. These include the fact that the development will be more attuned to the needs of the local residents. It will involve and build on the understanding generated by those who have the most intimate knowledge of the area, and it will engage those who are most affected by the development and decisions. Benefits, of a more inclusive system of planning accrue at three points: for children, for society and for the professionals involved.

This chapter has been entitled colliding worlds. On the road, collisions are avoided by the creation and imposition of an intricate network of rules, directives and designs, devised to facilitate the participation of all vehicles in the effective movement of people and goods. The process is based on mutual understandings held by all users of the transport system. It is based on principles of fairness, and on the recognition of the rights of all road users to access the system. To continue the road analogy, 'collisions' occur between children, young people, adults in general and planners in particular, because there is no agreed understanding. There is uncertainty about which direction the other is heading, and children and young people do not feel their interests are equally represented by the 'rules'. Children are like cyclists in many of our car-dominated cities. Cyclists are allowed to use the road network, but it is clear that their needs are secondary to those of motorized vehicles. In the event of a clash of interests, cyclists lose out. In planning for an environment that truly recognizes the needs and rights of children, we need to elevate children's status to that of trucks – clearly visible, a little noisy perhaps, but ignored to the detriment of all. Physical environments are particularly important for children, and it is in shaping the physical environment that planners can be especially influential and make a difference. Through the physical environment, children enter into wider physical and social worlds and make their way into the broader community. The physical environment provides planners and policy makers with opportunities to engage with children and young people in shaping the environment. To date, however, despite the presence of considerable enthusiasm among planners to embrace more child centred practice, children remain a low planning priority. It is unfortunate that they continue to lose out to other more pressing planning concerns.

References

Boyden, J. (1991) *Children of the Cities*, London: Zed Books Ltd.

Child Accident Prevention Trust (undated) *CAPT Fact Sheet*, Child pedestrians, CAPT website. Available online: www.capt.org.uk?FAQ/default.htm (accessed 15th June 2004).

Crane, P. and Dee, M. (2001) 'Young people public space and new urbanism', *Youth Studies Australia*, 20: 11–18.

Cunningham, C., Jones, M. and Barlow, M. (1996) *Town Planning and Children: A Case Study of Lismore, New South Wales, Australia*, Armidale NSW: Department of Geography and Planning, University of New England.

Davis, A. and Jones, L. (1996) 'The children's enclosure', *Town and Country Planning*, September: 233–5.

Driskell, D. (2002) *Creating Better Cities with Children and Youth*, Paris: UNESCO Publishing.
Engwicht, D. (1999) *Street Reclaiming: Creating Livable Streets and Vibrant Communities*, Annandale, Australia: Pluto Press.
Francis, M. (1988) 'Negotiating between children and adult design values in open space projects', *Design Studies*, 9 (2): 67–75.
Freeman, C. (1995) 'The changing nature of children's environmental experience: the shrinking realm of outdoor play', *The International Journal of Environmental Education and Information*, July–September, 14 (3): 259–81.
Freeman, C., Nairn, K. and Sligo, J. (2003) '"Professionalising" participation: from rhetoric to practice', *Children's Geographies*, 1 (1): 53–70.
Freeman, C., Aitken-Rose E. and Johnston, R. (2004) *Generating the Future? The State of Local Government Planning for Children and Young People in New Zealand*, Dunedin: Department of Geography, University of Otago.
Hil, R. and Bessant, J. (1999) 'SPACED-OUT? Young people's agency, resistance and public space', *Urban Policy and Research*, 17 (1): 41–9.
Hill, M. and Tisdall, K. (1997) *Children and Society*, London: Longman.
Holloway, S. L. and Valentine, G. (2000) 'Children's geographies: playing, living, learning', in Holloway, S. L. and Valentine, G. *Children's Geographies and the New Social Studies of Childhood*, New York: Routledge.
Institute for Public Policy (2002), *Making it Happen: Implementing New Zealand's Agenda for Children*, Auckland: Institute of Public Policy, AUT.
Kearns, R. A. and Collins, D. C. A. (2003) 'Crossing roads, crossing boundaries: empowerment and participation in a child pedestrian safety initiative', *Space and Polity*, 7 (2): 193–212.
Lennard, H. and Crowhurst-Lennard, S. H. (1992) 'Children in public places: some lessons from European cities', *Children's Environments*, 9 (2): 37–47.
Matthews, H. (2002) 'The street as a liminal space: the barbed spaces of childhood', in Christensen, P. and O'Brien, M. (eds) *Children in the City*, London: Routledge Falmer.
Melton, G. (2000) 'Parents and children: legal reform to facilitate children's participation', in Smith, A. B., Gollop, M., Marshall, K. and Nairn, K. (eds) *Advocating for Children*, Dunedin: University of Otago Press.
Ministry of Social Development (2002) *New Zealand's Agenda for Children: Making Life Better for Children*, Wellington: Ministry of Social Development.
Ministry of Social Development (2003) *Involving Children: A Guide to Engaging Children in Decision-making*, Wellington: Ministry of Social Development.
Ministry of Social Development (2004) *Toolkit for Child and Youth Participation in Local Government Decision-making Processes*, Wellington: Ministry of Social Development. Available online: www.msd.govt.nz/work-areas/children-and-young-people (accessed 2 September 2004).
Ministry of Youth Affairs (2002) *Youth Development Strategy Aotearoa*, Wellington: Ministry of Youth Affairs.
Ministry of Youth Affairs (2003) *'Keeping it Real': A Resource for Involving Young People; Youth Development Participation Guide*, Wellington: Ministry of Youth Affairs.
Morrow, V. (1999) 'We are people too: children's and young people's perspectives on children's rights and decision-making in England', *The International Journal of Children's Rights*, 7: 149–70.
Nairn, K., Panelli, R. and McCormack, J. (2003) 'Destabilising dualisms: young people's experiences of rural and urban environments', *Childhood*, 10: 9–42.
Panelli, R., Nairn, K. and McCormack, J. (2002) 'We make our own fun: reading the politics of youth with (in) the community', *Sociologia Ruralis*, 42 (2): 106–30.

Pooley, C. G., Turnbull, J. and Adams, M. (2004) *Changing Patterns of Everyday Mobility Full Report of Research Activities and Results*, Lancaster: Department of Geography, Lancaster University.

Residents' Association for Five Roads Home Zone (2005) *Five Roads Forum*. Available online: fiveroadsforum.org/_otherhomezones.htm (accessed 2 September 2004).

Roberts, I. and Power, C. (1996) 'Does the decline in child injury mortality vary by social class? A comparison of class specific mortality in 1981 and 1991', *British Medical Journal*, 313: 784–6.

Roberts, I., Norton, R. and Taua, B. (1996) 'Child pedestrian injury rates: the importance of "exposure to risk" relating to socioeconomic and ethnic differences, in Auckland, New Zealand', *Journal of Epidemiology and Community Health*, 50: 162–5.

Smith, A. B., Gollop, M., Marshall, K., and Nairn, K. (2000) *Advocating for Children*, Dunedin: University of Otago Press.

Smith, L. T., Smith, G. H., Boler, M. *et al.* (2002) '"Do you guys hate Aucklanders too?" Youth: voicing difference from the rural heartland', *Journal of Rural Studies*, 18 (2): 169–78.

Tranter, P. and Pawson, E. (2001) 'Children's access to local environments: a case-study of Christchurch, New Zealand', *Local Environment*, 6 (1): 27–48.

Tranter, P. J. and Whitelegg, J. (1994) 'Children's travel behaviours in Canberra: car dependent lifestyles in a low density city', *Journal of Transport Geography*, 2 (4): 265–73.

UNICEF (1992) *Convention on the Rights of the Child*, New York: United Nations Publications.

Valentine, G. (1996a) 'Children should be seen and not heard: the production and transgression of adult's public space', *Urban Geography*, 17 (3): 205–20.

Valentine, G. (1996b) 'Angels and devils: moral landscapes of childhood', *Environment and Planning D: Society and Space*, 14: 581–99.

Wazana, A., Kreuger, P., Raina, P. and Chambers L. (1997) 'A review of risk factors for child pedestrian injury rates: Are they modifiable?' *Injury Prevention*, 3: 295–304.

Wheway, R. and Millward, A. (1997) *Child's Play: Facilitating Play on Housing Estates*, York: Chartered Institute of Housing, UK.

White, R. (1996) 'No-go in fortress city: young people, inequality and space', *Urban Policy and Research*, 14 (1): 37–50.

White R, (2001) 'Youth participation in the design of public spaces', *Youth Studies Australia* 20: 19–26.

Williams, S. (2000) *Leisure Opportunities: Twizel Youth have a Say*, unpublished Masters thesis, Dunedin: University of Otago.

Woolley, H., Dunn, D., Spencer, C. *et al.* (1999) 'Children describe their experiences of the city centre: a qualitative study of the fears and concerns which may limit their full participation', *Landscape Research*, 24 (3): 287–301.

Young, M. and Wilmot, P. (1979) *Family and Kinship in East London*, Harmondsworth: Penguin Books.

Chapter 6

Children in the urban environment

A review of research

Neil Sipe, Nick Buchanan and Jago Dodson

Introduction

This chapter provides a concise review of scholarly attempts to understand the conditions and experiences of children in urban environments. The focus is on research undertaken by Western scholars that links the urban environment with the health, independence and well-being of children. The chapter is divided into three parts. The first reviews research undertaken prior to 1970. During this period scholarly research was primarily discipline-focused and it concentrated on the physical and mental health of children, and children's stepwise development and interaction with their surrounding environment. The second part of the chapter covers the 1970–90 period, which was distinguished by a concern with the complex themes of independence and children's comprehension and the response of their environment. Studies with attentiveness to such phenomenological themes were common in the 1970s. Since then the relationship between urban environments and children's health outcomes, as well as the complexity of the dynamic influence of social, cultural and economic factors on children's development, has become an increasing focus of scholarly attention. A key theme to emerge during these two decades was a shift towards more inter- and multi-disciplinary research. The third and final part of the chapter focuses on the body of research that has dominated the field since the early 1990s – the relationship between obesity, physical activity and urban environments.

Early studies of children and cities

From the mid nineteenth century onwards, the welfare of children in 'the city of the dreadful night' (Hall 2002) developed as a concern of scholars, governments and social commentators. While not specifically focused on children, Engels' (1968) catalogue of urban squalor and deprivation was among the first to present in stark detail both the physical and moral conditions of urban children as well as the public exclamations of concern expressed by socially concerned observers. Hall (2002: 18) quotes Mearns' (1883) pamphlet *The Bitter Cry of Outcast London*, which provoked substantial consternation in the UK through its reports of urban squalor:

> [t]he child-misery that one beholds is the most heart-rending and appalling element in these discoveries; and of these not least is the misery inherited from the vice of drunken and dissolute parents, and manifest in the stunted, misshapen, and often loathsome objects that we constantly meet in these localities.

Morality was a major concern of nineteenth-century social reformers with the activities of slum dwelling children causing particular anxiety for many observers and commentators. Housing conditions within the Victorian industrial city were identified as the root cause of much of misery and immorality. The 'one-room' system in which tenement housing was separated into single room dwellings, each sharing communal facilities, often containing a variety of petty manufacturing activities, such as rag-picking, sack-making and rabbit-pulling, was particularly notorious. The cramming of children into single-room dwellings with multiple adults was seen as especially deleterious to children's well-being. Hall (2002: 21) quotes Lord Shaftesbury's account of the effects of the one-room system of urban housing on children's behaviour:

> I will give an instance of the evil consequences of the one-room system, and this is not an instance of the worst kind. This case only happened last year, but it is of frequent occurrence. A friend of mine who is at the head of a large school, going down to one of the back courts saw two children of tender years, 10 or 11 years old, endeavouring to have sexual connection on the pathway. He ran and seized the lad and pulled him off, and the only remark of the lad was, '[w]hy do you take hold of me? There are a dozen of them at it down there'. You must perceive that this does not arise from sexual tendencies and that it must have been bred by imitation of what they saw.

Much public reform effort was subsequently expended in the late nineteenth century to ameliorate the impact of the social and environmental conditions of industrial urbanization on children's well-being. Dickens' literary depictions of urchins and poor children as apprentice criminals also contributed to a growing public concern with children's welfare and the potential detriment to society of unchecked growth in the number of child criminals.

A great deal of social advocacy in the UK in the nineteenth century surrounded the particular issue of child labour. The exploitation of children as a source of labour for urban manufacturing in the UK attracted ongoing political scrutiny. Successive Factory Acts – beginning with the 1802 Health and Morals of Apprentices Act – sought to regulate and limit the extent to which children were subjugated as labourers within urban industrial capitalism. Christian morality played a strong role in this legislative protection. Christian advocates recognized that children needed to be sufficiently schooled to ensure the inculcation of moral rectitude, an outcome unlikely to occur within the oppressive conditions of the sweathouse and slum. The achievement of broader nineteenth-century standards of social morality in turn depended on improving the conditions of urban children. However, beyond the moralistic concerns with children's welfare, and the anxieties about the future outcomes for a society in which deficient children became deficient adults, there were few serious attempts to comprehend the condition of urban children from a scholarly perspective that went beyond indignant descriptive accounts of poverty and deprivation. The comprehension of children as complex and socially vulnerable citizens within broader society was not well developed at this time. However, the descriptions of abject child misery that the social reformers had identified in their forays into urban slums in the nineteenth century provides a backdrop by which to compare subsequent and arguably more genteel scholarly engagement with the condition of urban children.

Systematic and socially theorized attempts to comprehend the experience and condition of children did not begin to emerge until the early twentieth century. One of the first attempts to systematically investigate and present material on urban children's welfare was at a 1911 exhibition in Chicago, which resulted in a handbook, *The Child in the City* (Bremner 1979). The exhibition used photographs, models and demonstrations to illustrate the plight of the 640,000 children then living in Chicago (Bremner 1979).

Other displays at the Chicago exhibition focused on the park, and playground development. This concern with ensuring the availability of urban playspaces for children was common to many of the urban planning debates at the time, including in the UK and Australia. This concern with planning for children was in part a reflection of the social position of women – advocacy on 'serious' planning issues was restricted to men – who took up the role of child advocates as a means of engaging with urban and environmental policy via their claim to some 'natural' expertise (Gatley 2005). Because playgrounds provided the necessary space for child development, they were seen by planners as an important design component of a proper modern neighbourhood. While there continued to be an emphasis on playground design in the engagement of the planning profession with children's concerns into the 1920s, this attentiveness preceded broader and more sophisticated recognition that play was essential for children's intellectual and social development (Frost 1989).

Concern with the psychological and social dimensions of children's play, including the significant role of children's interaction with the urban environment, had begun to grow. This concern formed a critical part of Perry's Neighbourhood Unit in the 1920s, and was also incorporated into Stein's designs of the late 1930s (Frank, Engelke and Schmid 2003).

Sociologists began to discover urban children in the 1930s. Reckless (1935) argued for three linked approaches to the study of children: (1) comparative studies to examine how children acquired different social skills in different environments; (2) studies investigating how children acquired developmental skills; and (3) studies looking at how children responded to intervention programmes.

In contrast to today's concern with child obesity, some research in the 1930s was concerned with *underweight* children. McNeill (1931) reported how all of the children starting the new school year in an American primary school were examined for physical defects, weighed and measured, and notices were sent to the parents of those who needed medical attention. The first several weeks of school consisted of talks on food, sleep, exercise and cleanliness in order to improve children's health. Comparable approaches were pursued in other countries; the UK, Australia and New Zealand all introduced 'school milk' programmes during the 1930s that aimed to improve children's health outcomes by ensuring a minimal nutritional intake. New Zealand still retains a residual system of health camps, which date from the establishment of the country's welfare state in 1935 (Tennant 1994). Historically, such camps sought to extract malnourished children from the squalid conditions of the urban slums and transport them temporarily to locations with perceived health promoting properties. The health camp locations were almost universally rural, and the programmes they operated subjected children to a regimen of nutritious diets, adventurous physical activities and education. These health camps were driven, however, by the contemporary welfare concerns of the 1930s, rather than based on scholarly assessments and prescriptions.

Sociological studies of urban children became methodologically more complex during the 1930s and 1940s, spurred in part by broader developments within this discipline. Children's behaviour relative to socio-economic conditions was studied by Bernard (1939). His findings showed that children in lower socio-economic groups were more likely to have a greater affiliation with their neighbourhood, and their peers within that neighbourhood, than those in higher socio-economic groups. The factor Bernard considered as the most likely cause of this result was the lower population density in the higher socio-economic group. Consequently, children in those areas had fewer possibilities or opportunities to make neighbourhood social contacts. Transportation was also identified by Bernard (1939) as playing a role in children's social well-being; family-related transport factors such as car ownership and the availability of adult escorts also contributed positively to children's social relationships.

Macdonald, Carson and Havighurst (1949) examined how children's leisure activities differed according to socio-economic background. They found systemic class differences in children's participation in organized leisure activities and in particular individual activities. Thus, few lower-class children participated in middle-class leisure activities, while upper-, middle- and lower-class children generally remained in their social strata. Macdonald noted that upper-class children were often sent to private schools, where they associated with other upper-class children often in a formal setting, while many middle-class children were sent to organized activities to learn the right manners, language and attitudes (Macdonald, Carson and Havighurst 1949: 505). Lower-class parents tended to leave their children to the street and alley playgroups of their neighbourhood.

Marcus and Moore's (1976) review argued that in the 1950s only one research effort – Lynch and Lukashok's (1956) investigation into childhood memories of the city – explored children's environments in a sufficiently critical way. Marcus and Moore (1976) stressed that scholarly evaluation of children's environments was a recent phenomenon.

In the United States in the early 1960s, the National Institute of Child Health and Human Development (NICHHD) was established within the National Institutes of Health (Aldrich 1979). It was set up to investigate social-behavioural development, including learning and biological processes. While research at the NICHHD began with a clinical team and a biologist, the project leaders quickly realized that architects, engineers and environmental planners were needed as collaborators (Aldrich 1979). This was one of the first examples of the application of an interdisciplinary approach being developed to comprehend the relationship between children's psychological and social well-being and urban environments.

It was not until the mid 1960s in the United Kingdom that the first studies began using empirical research to understand how children interacted with the urban environment. This government-funded research focused on children's play, incorporating playgrounds, roads and neighbourhood conditions within housing estates (Marcus and Moore 1976). The late 1960s also saw Lady Allen of Hurtwood's (1968) *Planning for Play*, which described what activities children did, where they preferred to do them and the design implications of their activities.

1970s to the 1990s: emergence of multidisciplinary approaches to research

Attention to the ways children perceived their surrounding urban environment developed in the 1970s, due to the establishment in 1968 of a United Nations Educational, Scientific and Cultural Organization (UNESCO) ten-year programme that aimed to increase the understanding of people-centred solutions to environmental problems (Chawla 2002). Kevin Lynch headed this interdisciplinary approach, which involved a combination of social researchers, natural scientists, architects, planners and other urban environmental practitioners. The aim of the initial project was to look at 'the way small groups of young adolescents use and value their spatial environment' (Lynch 1977: 1). The project was eventually named Growing Up In Cities and has been described as 'an international effort to understand young people's own perspectives on the urban environment' (Chawla and Malone 2003: 118). This research programme proved to be an important contribution to scholarly understandings of the condition of children within urban areas.

In contrast to much of the previous research on 'programmed' children's spaces, comprising urban environments that are deliberately and specifically constructed for use by children – such as parks, playgrounds or backyards – Lynch was interested in children's use of 'unprogrammed' space, comprising of local streets, courtyards and staircases where children would meet and play informal games (Lynch 1977). Other areas of interest to Lynch were children's 'time budgets': the amount of time children had to freely organize their own activities, and barriers to movement through the city, such as personal fear, dangerous traffic, a lack of spatial knowledge, the cost of public transport and parental controls. Part of the research involved asking children to draw maps of their neighbourhoods, which were found to differ depending on the environment children lived in (Lynch 1977). Lynch's efforts inspired other researchers to undertake comparable studies, including Moore and Young (1978) – who looked at children's home territories, place affiliation and pathways that were used to travel around the neighbourhood – and Ward (1979), who explored children's relationship with their urban environment.

Despite the growth of scholarly research examining children's health and the urban environment, Michelson and Roberts (1979) argued that other urban researchers in the late 1970s had failed to account for children and their urban surroundings. Examples of this alleged failure were Fischer's (1977) sociological treatise *The Urban Experience* and Cantor's (1977) *The Psychology of Place.* Both failed to recognize that adults and children have different perceptions of urban places. A study on the quality of life in new US suburbs by Zehner (1977) was also criticized by Michelson and Roberts (1979) for not sufficiently comprehending children's spatial experience of the world. Michelson and Roberts argued that Zehner's study concentrated on whether communities were a good place to raise children from an adult perspective and completely ignored the interrelationship between children and their urban environment.

Research on urban children, in the 1980s had paralleled themes raised in the previous decade. The emphasis during this period was on children in the physical urban environment and how the environment had mental and social effects on the development of children. As in the 1970s, such research in this era concentrated on concerns such as how land uses influenced children's play patterns, the impact of the physical environment on access to playspace and children's social interaction, and how safety and mobility affected play (Berg and Medrich 1980). Berg and Medrich made two important observations in relation to safety and mobility: first, that urban spaces with poor physical accessibility had fewer spontaneous activities than areas with high accessibility;

second, that children sought out unplanned playspaces where they were free to explore the area and invent their own activities. This was consistent with earlier findings from Lynch (1977). Such areas were termed the 'fourth environment'. The recognition of the importance of unregulated children's playspaces prompted research looking at children beyond conventional play areas, such as parks and backyards, to include streets, shops, clinics, cinemas, buses, offices and cafes (Van Vliet 1983).

Attempts were also made in the 1980s to incorporate children's ideas and perspectives about their neighbourhoods into policy and research. De Monchaux (1981) argued that children's needs in the urban environment have historically been assumed rather than proved, and that decision makers need to have a greater understanding of the factors that are important in children's engagement with the environment. Later research (Homel and Burns 1985) looked at environmental differences for inner-city suburban children and those living in more privileged environments, and attempted to understand how children perceived the quality of their surrounding environment. By the end of the 1980s, more complex statistical approaches were developed (Homel and Burns 1989) to define environmental quality and accordingly to comprehend the well-being of children using family characteristics as controlling variables.

While accessibility issues for children had been raised in the 1970s and 1980s, these became increasingly important in the early 1990s. Cahill (1990), for example, argued that American children's access to public places was severely restricted. Gaster (1991) argued that children and youth in cities had progressively been cut off from safely using and enjoying their neighbourhoods since the 1970s. Gaster identified increasing crime, automobile traffic and the deterioration or destruction of parks, playgrounds and schoolyards as factors contributing to this change. In Australia, Tranter and Whitelegg (1994) have examined the independent mobility of children (the freedom to travel around their own neighbourhood or city without adult supervision) while McMeeking and Purkayastha (1995) examined the problems of accessibility and mobility encountered by young adolescents as they pursued leisure activities.

1990s and beyond: obesity, physical activity and the urban environment

Physical health and medical issues concerning urban children emerged as a research area in the early 1990s. Kuntzleman (1993) reported controversy surrounding the physical fitness levels of American youth. Some experts claimed that American youth were the most active and fit age group in the country while others argued that there had been an erosion of youth fitness levels during the previous two to three decades. International comparative research found that children in the United States were fatter, slower and weaker than their counterparts in other developed nations as a result of a decline in physical fitness programmes in schools and an increase in sedentary activities, such as watching television (DiNubile 1993). Adverse health effects in children from exposure to dangerous pollutants became more prominent in the 1990s, as pollutants in the urban environment were found to be present in children (Olden 1993).

The proliferation of research since the mid 1990s provides evidence of the growing emphasis by several disciplines on children's health. Since the turn of the millennium, there has been a deluge of research on children's health, particularly physical health. Many disciplines have become involved as they have seen their interests entwined with others across what has become a wide research field. The urban planning and transportation literature has investigated the relationship between the built environment and transportation mode choice, including walking and bicycling (Cervero and Duncan 2003). Research in medical and public health journals advocates increased walking and bicycling as good forms of physical activity to improve public health (Dora 1999). Public health practitioners have also begun working with researchers in urban planning and related fields (Frank, Anderson and Schmid 2004).

Since the mid 1990s there has been a large increase in the number of journals dedicated to children's health and the urban environment. This increase is due, in part, to other disciplines becoming involved in children's health, thus broadening the field of inquiry and stimulating demand for multidisciplinary venues for publication. In the 1970s and 1980s, research in this area was conducted primarily by social scientists, natural scientists, architects, planners, psychologists and other environmental practitioners. While these disciplines remain interested in children's health, researchers from other disciplines have begun to take an interest, including health, medicine, nutrition, physical education and sports science. Recent publications both reflect and fuel interest in this topic. A number of journals have devoted special issues to the built environment and health, including: *The American Journal of Preventive Medicine* (May 2002), *The Institute for Transportation Journal* (June 2003), *Landscape and Urban Planning* (September 2003) and *The American Journal of Public Health* (September 2003).

Killingsworth, Earp and Moore (2003) argue that this renewed interest in children's urban experiences and outcomes has revitalized the call to embrace interdisciplinary collaboration and to identify ways to build a research agenda to address the new demand for knowledge. Multidisciplinary approaches have evolved specifically to confront issues of children's health (King *et al.* 2002; Frumkin, Frank and Jackson 2004). Much of the urban child health research in the late 1990s focused on identifying the underlying causes of inactivity and obesity among children in the Western world (Grundy 1998; Katzmarzyk *et al.* 1998; Kohl and Hobbs 1998). However, there were other streams of research during this period dealing with: transport and disadvantaged groups (including children) – Morgan and Robbins (1997), and Tolley (1997); children as a disadvantaged group in society – Malone and Hasluck (1998), Millward (1998), and Bartlett *et al.* (1999); and parental concerns in selecting playspaces for their children – Sallis *et al.* (1997).

As an outcome of research efforts in the 1990s and increasing attention from the media (Powell 2000; Sommerfeld 2005), recent research has attempted to determine the causes of rising child obesity. For example, in the United States, 16 per cent of children are obese and a further 15 per cent are likely to become obese (Hedley *et al.* 2004; Neumark-Sztainer 2005). These trends have resulted in numerous publications on how the built environment is affecting children's health (for example Cummins and Jackson 2001; LeClair 2001; Handy *et al.* 2002; Giles-Corti and Donovan 2002; Dannenberg *et al.* 2003; Frank, Engelke and Schmid 2003; Jackson 2003; Killingsworth, Earp and Moore 2003; Crowhurst-Lennard and Riley 2004; Frumkin, Frank and Jackson 2004).

A majority of researchers on children's health and the built environment are US-based, and several have placed the blame for declining children's health on urban sprawl (Frumkin 2002; Burchell and Mukherji 2003; Ewing *et al.* 2003; Krizek, Birnbaum and Levinson 2004; Sturm and Cohen 2004). However, there has also been growing concern in Australia (Catford and Caterson 2003) and the United Kingdom (Lumsden and Mitchell 1999) about overweight and obese children. Catford and Caterson (2003) found that childhood obesity levels in Australia are comparable to the US and exceed those in the United Kingdom. They attribute this pattern to a decline in physical activity and a rise in nutritional energy intake. Frank, Engelke and Schmid (2003) have argued that high rates of childhood obesity suggest a link between environmental conditions, personal behaviour and body weight. One of the most important factors identified in Australia is that children are not undertaking sufficient physical activity, thus exacerbating their risk of becoming obese (Waters and Baur 2003).

While the impact of the urban environment on health has yet to be fully explored, Krizek (2003: 28) suggests that:

> there is considerable enthusiasm among individuals in research, advocacy and policy circles for the idea that 'good' urban design will positively contribute to levels of physical activity . . . most agree it is critically important to support planning efforts that make physical activity and 'active travel' easy, available to diverse and increased populations and more attractive.

However, Krizek warns that health-sensitive urban design by itself may not necessarily lead to better physical health, and the independent effect of urban design on physical activity may be less significant once other issues have been accounted for. Urban form is viewed by many researchers as only one variable that contributes to adverse health outcomes. Thus, for example, Frumkin *et al.* (2004:

190) acknowledge that sprawl 'does not fully explain the inactivity and overweight that plague American children'. Other possibilities Frumkin Frank and Jackson cite as contributing to the increase in physical inactivity include: (1) dietary changes with 'super-sized' portions of food; (2) behavioural shifts towards sedentary leisure activities with more television and computer time; and (3) other factors such as reductions in school-based physical education programmes. While the physical environment surrounding the child is an important health factor, Lobstein, Baur and Uauy (2004) argue that it is also important to consider the micro-environment created in the home. For younger children, the family environment plays an important role in determining their risk of obesity. Parental physical activity levels, the family's eating behaviour and television viewing habits all contribute to problems of overweight and obesity (Lobstein, Baur and Uauy 2004).

Many studies have sought to identify the cause(s) of increasing rates of childhood obesity in Western countries. While much of the research has examined the relationship between the built environment and health, several other interrelated factors have also been studied. Cummins and Jackson (2001) argued that the causes of the current obesity epidemic are complex, with inadequate physical activity being a crucial factor. Physical activity declines when children lack sufficient opportunities to exercise during or after school and when they rely on private automobile transportation rather than walking, biking or public transport (Grundy 1998). Collins and Kearns (2001) believe that the retreat by children into the car, and indoors, has been motivated in large part by adults' concerns for their safety, and has increasingly been associated with a broader societal shift towards sedentary lifestyles, declining fitness and weight problems. Lobstein, Baur and Uauy (2004) have argued that the greatest health problems arising from insufficient physical activity will be apparent in the next generation as today's obese children become adults and begin to suffer chronic obesity-related health problems.

The focus on the obesity 'epidemic' (Nestle and Jacobson 2000) or 'pandemic' (Kimm and Obarzanek 2002) has resulted in a shift away from the cognitive and mental development research that was prevalent in the 1970s. Those studies focused on how children perceived their surrounding environment, and examined how children's cognitive abilities developed as a result of their environment. In the late 1990s, Sallis *et al.* (1997) found that children's play promotes social and cognitive development and provides physical activity that has health benefits. Driskell (2002: 22) states that the local environment can 'pose significant threats to young people's physical, mental, emotional and social development, or it can provide positive development opportunities for them to explore, grow and engage with the world'. Some studies have looked at perceived neighbourhood attributes affecting physical activity, although children have not been a focus (Saelens *et al.* 2003; Humpel *et al.* 2004; Owen *et al.* 2004).

Christensen and O'Brien (2003) discuss the mental development of children. This dimension still remains a relevant area of inquiry as the urban environment allows children to develop social competencies and to enact their growing maturity through independent actions in and away from their home and neighbourhood. With children being able to explore their local area and neighbourhood, they come to build an understanding of their local place, and are able to become independent movers within the urban environment (Christensen 2003). With an increase in parental concerns for child safety, more and more children are being kept indoors (Collins and Kearns 2001). This trend towards restrictive supervision impacts on how children learn about their immediate environment, and could also impact on how children increase their home range as they grow up (Christensen and O'Brien 2003). The loss of unstructured urban space has also contributed to this trend (McKendrick, Bradford and Fielder 2000; Rasmussen and Smidt 2003; Zeiher 2003). Other studies, such as Tranter and Pawson (2001), have focused on access to local environments. Tranter and Pawson incorporated a social aspect in addition to the physical environment, and also addressed the issue of child independence, while Evans (2003) looked at links between the built environment and mental health, although children were barely mentioned in the findings.

Driskell (2002) has argued that young people should be included in community development processes. This is because children are intimately familiar with the local environment, and are often the most knowledgeable on how the local environment and development decisions will impact on their own lives and that of their community. Driskell also argues that children's participation in community development can be beneficial for both children – as they learn new skills – and adults – as they develop a better understanding of children's perspectives of the local environment. Driskell's work shares similarities with that of Christensen and O'Brien (2003), who argue that their book *Children in the City: Home, Neighbourhood and Community* has three main themes. The first theme is inclusivity, where attention needs to be paid to children both as a social group and as individuals, and understanding the city from children's perspectives. The second theme is about the myriad of overlapping relationships that children are exposed to in the city. The third theme is that children also need to be involved in the process of change in cities. Young people's participation and representation in society has also been identified as an important factor by Matthews, Limb and Taylor (1999).

However, despite the growing research on the relationship between health and the urban environment, there remains doubt as to the exact nature of this relationship. The US Transport Research Board (TRB) (2005) has recently completed a study that reviews the trends affecting the relationships between physical activity, health, transportation and land use. While not concentrating specifically on children, it does reveal some deficiencies in the current research. The study argues that the built environment is one of many variables affecting physical activity levels and that much remains to be learned about the importance of the individual (e.g. physical capacity, attitudes, preferences and time demands), the social context (e.g. social norms and support networks) and the physical environment as determinants of physically active behaviour. The diversity and complexity of the urban environment appears to be a critical factor impeding the scholarly disentanglement of the multiple variables contributing to children's, and other groups', health and socio-economic outcomes.

Since research on health and the urban environment is relatively new and still developing, one of the key findings of the TRB review is the observation that the field 'is not sufficiently advanced to support causal connections or to identify with certainty those characteristics of the built environment most closely associated with physical activity behaviour' (TRB 2005: 219). Another key outcome of the TRB report was that strong and continuing research effort is needed to further understand the relationship between the built environment and physical activity. The final conclusion from the report was that 'modifications to the built environment alone are unlikely to solve the public health problem of insufficient physical activity' (TRB 2005: 220) and that complementary strategies addressing individual, social and environmental determinants of physical activity need to be developed.

Conclusion

Research on children's health and the urban environment has changed substantially, especially over the last ten years. The moral panics of the nineteenth century and the beginnings of attentive scholarly studies of children's developmental patterns and mental health have made way for a contemporary agenda increasingly focused on identifying links between the urban environment and children's physiological health, specifically obesity.

Prior to 1930, the focus of much research was on children's playgrounds and creating nurturing urban environments through the provision of parks, playgrounds and other facilities. Playgrounds were seen to provide a necessary space for children to develop physically, emotionally and socially with other children. Sociological interest in children within the urban environment developed in the mid 1930s, with several publications dealing with issues of acquiring skills and knowledge and whether these processes varied by socio-economic status.

Research examining the relationship between urban children and their environment, including effects on mental development, independence and interactions and perceptions of surrounding environments, did not begin to develop until the 1960s, and it was not until the 1970s that major research efforts gained scholarly momentum. Researchers in the United Kingdom appear to have been more attentive to research questions relating to urban children compared to researchers in the United States, with studies of children in urban environments occurring in the UK earlier than in the US. The late 1970s was typified by numerous global efforts (Lynch 1977; Ward 1979) to understand how children interacted with the urban environment. The main change in the 1980s was the attempt to incorporate children's ideas and perspectives into policy documents.

Since the mid 1990s there has been a dramatic shift in research towards children's health and the urban environment. There has been a move away from research on children's mental health to one fixated on children's physical health and problems with inactivity and childhood obesity. The number of books and articles cited in this chapter emphasize how dominant within scholarly and policy thinking the issue of child obesity has become. Researchers in the field have stressed the need for multidisciplinary approaches to combat obesity including urban planning, transport, and public health and medicine. Interdisciplinary approaches are not new to the wider field of children's health and the urban environment, as evidenced from collaborative efforts since the 1970s. Another theme to emerge from the current research is the need for increased children's participation in planning and decision-making matters. While much of the recent attention has been on the links between child obesity and the urban environment, some work has continued to assess children's perceptions of, and experiences with, their surrounding environment. As a precautionary note, the final theme from this chapter is that the link between the urban environment, physical activity and childhood obesity has yet to be conclusively established, and therefore warrants further attention.

References

Aldrich, R. (1979) 'The influences of man-built environment on children and youth', in Michelson, W., Levine, S. and Michelson, E. (eds) *The Child in the City: Today and Tomorrow*, The Child in the City Programme, Toronto: University of Toronto Press.

Bartlett, S., Hart, R., Satterthwaite, D., de la Barra, X. and Missair, A. (1999) *Cities for Children: Children's Rights, Poverty and Urban Management*, London: Earthscan Publications.

Berg, M. and Medrich, E. (1980) 'Children in four neighbourhoods: the physical environment and its effect on play and play patterns', *Environment and Behavior*, 12 (3): 320–48.

Bernard, J. (1939) 'The neighborhood behavior of school children in relation to age and socioeconomic status', *American Sociological Review*, 4 (5): 652–62.

Berrigan, D. and Troiano, R. (2002) 'The association between urban form and physical activity in US adults', *American Journal of Preventive Medicine*, 23 (2S): 74–9.

Bremner, R. (1979) 'The child in the city: continuity and change in problems and programmes since 1875', in Michelson, W., Levine, S. and Michelson, E. (eds) *The Child in the City:*

Today and Tomorrow, The Child in the City Programme. Toronto: University of Toronto Press.

Burchell, R. and Mukherji, S. (2003). 'Conventional development versus managed growth: the costs of sprawl', *American Journal of Public Health,* 93 (9): 1534–40.

Cahill, S. (1990). 'Childhood and public life: reaffirming biographical divisions', *Social Problems,* 37 (3): 390–402.

Cantor, D. (1977). *The Psychology of Place*, London: Architectural Press.

Catford, J. and Caterson, I. (2003) 'Snowballing obesity: Australians will get run over if they just sit there', *Medical Journal of Australia*, 179: 577–9.

Cervero, R. and Duncan, M. (2003) 'Walking, bicycling, and urban landscapes: evidence from the San Francisco Bay area', *American Journal of Public Health,* 93 (9): 1478–83.

Chawla, L. (2002) 'Cities for human development', in Chawla, L. (ed.) *Growing Up in an Urbanizing World,* Paris: UNESCO Publishing/Earthscan Publications.

Chawla, L. and Malone, K. (2003) 'Neighbourhood quality in children's eyes', in Christensen, P. and O'Brien, M. (2003) *Children in the City: Home, Neighbourhood and Community.* London: Routledge/Falmer.

Christensen, P. (2003) 'Place, space and knowledge: children in the village and the city' in Christensen, P. and O'Brien, M. (eds) *Children in the City: Home, Neighbourhood and Community*, London: Routledge/Falmer.

Christensen, P. and O'Brien, M. (2003) 'Children in the city: introducing new perspectives', in Christensen, P. and O'Brien, M. (eds) *Children in the City: Home, Neighbourhood and Community,* London: Routledge/Falmer.

Collins, D. and Kearns, R. (2001) 'The safe journeys of an enterprising school: negotiating landscapes of opportunity and risk', *Health and Place*, 7: 293–306.

Crowhurst Lennard, S. and Riley, J. (2004) 'Children and the built environment', *Urban Land*, 63 (1): 69.

Cummins, S. and Jackson, R. (2001) 'The built environment and children's health', *Pediatrics Clinics of North America,* 48 (5): 1241–52.

Dannenberg, A., Jackson, R., Frumkin, H., Schieber, R., Pratt, M., Kochtitzky, C. and Tilson, H. (2003) 'The impact of community design and land-use choices on public health: a scientific research agenda', *American Journal of Public Health*, 93 (9): 1500–8.

DiNubile, N. (1993) 'Youth fitness – problems and solutions', *Preventive Medicine,* 22: 589–94.

Dora, C. (1999) 'A different route to health: implications of transport policies', *British Medical Journal,* 318 (7199): 1686–9.

Driskell, D. (2002) *Creating Better Cities with Children and Youth: A Manual for Participation*, Paris: UNESCO Publishing/Earthscan Publications.

Engels, F. (1968) *The Condition of the Working Class in England*, Palo Alto, CA: Stanford University Press (trans: W. O. Henderson and W. H. Chaloner).

Evans, G. (2003) 'The built environment and mental health', *Journal of Urban Health: Bulletin of the New York Academy of Medicine,* 8(4): 536–55.

Ewing, R., Schmid, T., Killingsworth, R., Zlot, A. and Raudenbush, S. (2003) 'Relationship between urban sprawl and physical activity, obesity, and morbidity', *American Journal of Health Promotion,* 18 (1): 47–57.

Fischer, C. (1977) *The Urban Experience,* New York: Harcourt, Brace and Jovanovich.

Frank, L., Engelke, P. and Schmid, T. (2003) *Health and Community Design: The Impact of the Built Environment on Physical Activity*, Washington, DC: Island Press.

Frank, L., Anderson, M. and Schmid, T. (2004) 'Obesity relationships with community design, physical activity and time spent in cars', *American Journal of Preventive Medicine,* 27 (2): 87–96.

Frost, J. (1989) 'Play environments for young children in the USA: 1800–1990', *Children's Environments Quarterly*, 6 (4): 17–24.

Frumkin, H. (2002) 'Urban sprawl and public health', *Public Health Reports*, 117: 201–17.

Frumkin, H., Frank, L. and Jackson, R. (2004) *Urban Sprawl and Public Health: Designing, Planning and Building for Healthy Communities*, Washington, DC: Island Press.

Gaster, S. (1991) 'Urban children's access to their neighborhood: changes over three generations', *Environment and Behavior*, 23 (1): 70–85.

Gatley, J. (2005) 'For king and empire: Australian women and nascent town planning', *Planning Perspectives*, 20 (2): 121–45.

Giles-Corti, B. and Donovan, R. (2002) 'The relative influence of individual, social and physical environment determinants of physical activity', *Social Science & Medicine*, 54: 1793–1812.

Grundy, S. (1998) 'Multifactorial causation of obesity: implications for prevention', *American Journal of Clinical Nutrition*, (67) (suppl): 563S-572S.

Hall, P. (2002) *Cities of Tomorrow*, 3rd edn, Oxford: Blackwell Publishing.

Handy, S., Boarnet M., Ewing, R. and Killingsworth, R. (2002) 'How the built environment affects physical activity: views from urban planning', *American Journal of Preventive Medicine*, 23 (2S): 64–73.

Hedley, A., Ogden, C., Johnson, C., Carroll, M., Curtin, L. and Flegal, K. (2004) 'Prevalence of overweight and obesity among US children, adolescents and adults, 1999–2002', *Journal of the American Medical Association*, 291 (3): 2874–50.

Homel, R. and Burns, A. (1985) 'Through a child's eyes: quality of neighbourhood and quality of life', in Burnley, I. and Forrest, J. *Living in Cities: Urbanism and Society in Metropolitan Australia*. Sydney: Allen & Unwin.

Homel, R. and Burns, A. (1989) 'Environmental quality and the well-being of children', *Social Indicators Research*, 21: 133–58.

Humpel, N., Owen, N., Iverson, D., Leslie, E. and Bauman, A. (2004) 'Perceived environmental attributes, residential location, and walking for particular purposes', *American Journal of Preventive Medicine*, 26 (2): 119–25.

Jackson, R. (2003) 'The impact of the built environment on health: an emerging field', *American Journal of Public Health*, 93 (9): 1382–4.

Katzmarzyk, P., Malina, R., Song, T. and Bouchard, C. (1998) 'Television viewing, physical activity, and health-related fitness of youth in the Quebec family study', *Journal of Adolescent Health*, 23: 318–25.

Killingsworth, R., Earp, J. and Moore, R. (2003) 'Supporting health through design: challenges and opportunities', *American Journal of Health Promotion*. 18 (1): 1–2.

Kimm, S. and Obarzanek, E. (2002) 'Childhood obesity: a new pandemic of the new millennium', *Pediatrics*, 110 (5): 1003–7.

King, A., Stokols, D., Talen, E., Brassington, G. and Killingsworth, R. (2002) 'Theoretical approaches to the promotion of physical health: forging a transdisciplinary paradigm', *American Journal of Preventive Medicine*, 23 (2S): 15–25.

Kohl, H. and Hobbs, K. (1998) 'Development of physical activity behaviors among children and adolescents', *Pediatrics*, 110 (Supplement): 549–54.

Krizek, K. (2003) 'The complex role of urban design and theoretical models of physical activity', *Progressive Planning*, No. 157, Fall 2003: 28–9.

Krizek, K., Birnbaum, A. and Levinson, D. (2004) 'A schematic for focusing on youth in investigations of community design and physical activity', *American Journal of Health Promotion*, 9: 33–8.

Kuntzleman, C. (1993) 'Childhood fitness: what is happening? what needs to be done?', *Preventive Medicine*, 22: 520–32.
Lady Allen of Hurtwood (1968) *Planning for Play*, London: Thames & Hudson; later published by MIT Press.
LeClair, J. (2001) 'Children's behaviour and the urban environment: an ecological analysis', *Social Science & Medicine*, 53: 277–92.
Lobstein, T., Baur, L. and Uauy, R. (2004) 'Obesity in children and young people: a crisis in public health', *Obesity*, 5 (Supplement 1): 4–85.
Lumsden, L. and Mitchell, J. (1999) 'Walking, transport and health: do we have the right prescription?', *Health Promotion International*, 14 (3): 271–9.
Lynch, K. (1977) *Growing Up in Cities*, Cambridge, MA: MIT Press.
Lynch, K. and Lukashok, A. (1956) 'Some childhood memories of the city', *Journal of the American Institute of Planners*, Summer.
Macdonald, M., Carson, M. and Havighurst, R. (1949) 'Leisure activities and the socio-economic status of children', *The American Journal of Sociology*, 54 (6): 505–19.
McKendrick, J., Bradford, M. and Fielder, A. (2000) 'Kid customer? commercialization of playspace and commodification of childhood', *Childhood*, 7 (3): 295–314.
McMeeking, D. and Purkayastha, B. (1995) '"I can't have my mom running me everywhere": adolescents, leisure, and accessibility', *Journal of Leisure Research*, 27 (4): 360–78.
McNeill, N. (1931) 'Health and safety project', *Journal of Educational Sociology*, 5 (4): 240–2.
Malone, K. and Hasluck, L. (1998) 'Geographies of exclusion: young people's perceptions and use of public space', *Family Matters*, 49: 20–6.
Marcus, C. and Moore, R. (1976) 'Children and their environments: a review of research 1955–1975', *Journal of Architectural Education*, 29 (4): 22–5.
Matthews, H., Limb, M. and Taylor, M. (1999) 'Young people's participation and representation in society', *Geoforum*, 30 (2): 135–44.
Mearns, A. (1883) *The Bitter Cry of Outcast London: An Inquiry into the Condition of the Abject Poor*, London: James Clarke.
Michelson, W. and Roberts, E. (1979) 'Children and the urban physical environment', in Michelson, W., Levine, S. and Spina, A. (eds) *The Child in the City: Changes and Challenges*, Child in the City Programme, Toronto: University of Toronto Press.
Millward, A. (1998) 'Seen – but not yet heard', *Town and Country Planning*. 67: 354–7.
Monchaux, S. de (1981) *Planning with Children in Mind: A Notebook for Local Planners and Policy Makers on Children in the City Environment*, Sydney: NSW Department of Environment and Planning.
Moore, R. and Young, D. (1978) 'Childhood outdoors: toward a social ecology of the landscape', in Altman, I. and Wohlwill, J. *Children and the Environment*, New York: Plenum Press.
Morgan, D. and Robins, S. (eds) (1997) *Road Transport and Health*, London: British Medical Association.
Nestle, M. and Jacobson, M. (2000) 'Halting the obesity epidemic: a public health policy approach', *Public Health Reports*, 115: 12–24.
Neumark-Sztainer, D. (2005) 'Addressing obesity and other weight related problems in youth', *Archives of Pediatrics and Adolescent Medicine*, 159 (3): 290–1.
Olden, K. (1993) 'Environmental risks to the health of American children', *Preventive Medicine*, 22: 576–8.
Owen, N., Humpel, N., Leslie, E., Bauman, A. and Sallis, J. (2004) 'Understanding environmental influences on walking: review and research agenda', *American Journal of Preventive Medicine*, 27 (1): 67–76.

Powell, S. (2000) 'One in four Australian children is overweight: slower, stiffer, heavier – they are the cotton-wool generation', *The Weekend Australian* (Review Section), 27–8 May 2000: 6–8.

Rasmussen, K. and Smidt, S. (2003) 'Children in the neighbourhood: the neighbourhood in the children', in Christensen, P. and O'Brien, M. (eds) *Children in the City: Home, Neighbourhood and Community*. London: Routledge/Falmer.

Reckless, W. (1935) 'As sociologists enter child-development study', *Journal of Educational Sociology'* 9 (2): 111–18.

Saelens, B., Sallis, J., Black, J. and Chen, D. (2003) 'Neighborhood based differences in physical activity: an environment scale evaluation', *American Journal of Public Health*, 93 (3): 1552–8.

Sallis, J., McKenzie, T., Elder, J., Broyles, S. and Nade, P. (1997) 'Factors parents use in selecting play spaces for young children', *Archives of Pediatrics and Adolescent Medicine*, 151 (4): 414–17.

Sommerfeld, J. (2005) 'Fat kids eating way to early death', *The Courier-Mail*, Brisbane, 24 March: 1.

Sturm, R. and Cohen, D. (2004) 'Suburban sprawl and physical and mental health', *Public Health*, 118: 488–96.

Tennant, M. (1994) *Children's Health the Nation's Wealth: A History of Children's Health Camps*, Wellington: Bridget Williams Books.

Tolley, R. (ed.) (1997) *The Greening of Urban Transport: Planning for Walking and Cycling in Western Cities*, West Sussex: John Wiley & Sons.

Transport Research Board (TRB) (2005) *Special Report 282: Does the Built Environment Influence Physical Activity? Examining the Evidence*, Washington, DC: National Academy of Sciences.

Tranter, P. and Whitelegg, J. (1994) 'Children's travel behaviours in Canberra: car-dependent lifestyles in a low-density city', *Journal of Transport Geography*, 2 (4): 265–73.

Tranter, P. and Pawson, E. (2001) 'Children's access to local environments: a case study of Christchurch, New Zealand', *Local Environment*, 6 (1): 27–48.

Van Vliet, W. (1983) 'Exploring the fourth environment: an examination of the home range of city and suburban teenagers', *Environment and Behavior*, 15 (5): 567–88.

Ward, C. (1979) *The Child in the City*, Harmondsworth: Penguin Books.

Waters, E. and Baur, L. (2003) 'Childhood obesity: modernity's scourge', *Medical Journal of Australia*, 178: 422–3.

Zehner, R. (1977) *Indicators of the Quality of Life in New Communities.* Chapel Hill, NC: University of North Carolina.

Zeiher, H. (2003) 'Shaping daily life in urban environments', in Christensen, P. and O'Brien, M. (eds) *Children in the City: Home, Neighbourhood and Community*. London: Routledge/Falmer.

Part Three
Spheres of action

Chapter 7

Children in the intensifying city

Lessons from Auckland's walking school buses

Robin Kearns and Damian Collins

Introduction

In a recent critique of Australian urbanism, Gleeson (2004) notes the demise of the conventional low-density suburb and the emergence of a bifurcated landscape consisting of spaces of intense poverty and despair on the one hand, and elite master-planned estates filled with mega-houses on the other. Both environments have negative implications for children's freedom of movement. In the former, the presence of crime, prostitution and drug-dealing contributes to a public environment that is often antithetical to safe and happy childhoods. In the latter, risk-averse parenting cultures mean children are likely to be chauffeured to and from extra-curricular activities, and school, by private motor vehicle. In this chapter, we seek to link concerns about children's increasing physical inactivity and confinement to private supervised spaces to the increasingly intense nature of urban and suburban landscapes.

In the larger cities of Australia and New Zealand, backyards, fields and wooded areas are increasingly clear-cut, filled-in and paved-over – replaced with impermeable structures in response to population growth, demand for higher-density housing and an awareness of the costs of sprawl. In combination with underdeveloped public transport networks, one consequence is a marked increase in traffic volumes and congestion, a trend that has contributed to parental fears for children's safety in public space, as well as to chauffeuring behaviour. There is a pressing need for a reworking of urban space that encompasses new, inclusive civic forms that agitate against trends towards social segregation and privatism. In this chapter, we consider the walking school bus (WSB) as a promising, although partial, response to the challenges posed by urban intensification in Auckland, New Zealand.

The WSB phenomenon, and the Auckland region, are introduced in the following section, which also provides an international context for our research, noting relevant trends in both urban intensification and children's well-being. After briefly reviewing the methods employed in our study, results from a survey of 42 WSB coordinators are presented. We observe that this volunteer-driven initiative has been most widely adopted in relatively privileged neighbourhoods, where the public health challenges associated with such problems as child pedestrian injury and childhood overweight/obesity are generally least pressing. Moreover, the perceived benefits of WSB initiatives can be closely linked to the problems associated with intensification in the Auckland region. The penultimate section considers the long-term viability of WSBs, and identifies several 'lessons' that can be drawn from their success to date. The concluding section reflects on the child's place in a city that continues to be dominated by automobiles.

Context

Within the Auckland urban area (population 1.3 million – 2003 estimate), a high level of car dependence, an underdeveloped public transport system and residential intensification have exacerbated traffic congestion and parental chauffeuring behaviour. The former has become the dominant issue in municipal politics, with most attention focused on the economic costs of congestion, the planned construction of new motorways, and – to a lesser extent – the provision of dedicated lanes for motorized buses. Within this milieu, the walking school bus has emerged as a popular option for reducing the congestion associated with the school run, and increasing children's physical activity. By tapping into latent demand for active travel on the part of many primary school pupils (see O'Brien 2001; Mitchell 2005) WSBs may help to break the 'social trap' of ever-increasing reliance on the private automobile for everyday journeys (see Tranter and Pawson 2001; also Chapter 8 in this volume).

The WSB involves groups of children walking to and from school under adult supervision along a set route complete with specified stops at which they may embark or disembark. Adult volunteers – usually parents – guide the bus between its farthest stop and the school, while informally maintaining discipline and remaining alert for potential hazards and obstacles (see Figures 7.1 and 7.2). By offering a safe and reliable alternative to car travel, the WSB may re-legitimize the act of children walking on city streets. While not a substitute for the unsupervised active travel that children traditionally undertook in large numbers (Kearns, Collins and Neuwelt 2003), the WSB has emerged at a time when other opportunities for outdoor exercise are being curtailed by urban intensification and parents' responses to it.

The term 'urban intensification' is commonly used to refer to a range of processes that make cities more compact, including the consolidation of population and dwellings within existing urban areas (Williams, Burton and Jenks 1996). These processes produce higher densities of people, activities and buildings, and are commonly employed in efforts to reduce suburban sprawl and greenfield development. Intensification has the potential to reduce automobile dependency, by increasing the viability of public transit services, and enabling more people to live within walking or cycling distance of workplaces. However,

7.1 Negotiating a pedestrian crossing.

Source: Sue Kendall, Walking School Bus coordinator, Auckland Region.

7.2 Mass transit on the walking school bus.

Source: Sue Kendall, Walking School Bus coordinator, Auckland Region.

this benefit is commonly unrealized (Troy 1996). When increasing densities are not accompanied by reduced car dependence, the result is an increase in congestion, and this factor – combined with the potential loss of both public and private green space to subdivision and new construction (Williams, Burton and Jenks 1996) – may detract significantly from the quality of the urban environment. Thus, one significant outcome of intensification may be a city that is increasingly hostile to children's needs.

The population of the Auckland region is anticipated to reach 2.2 million people by 2050, with 30 per cent of that number living in high-density housing (Lyne and Moore 2004). The move to terraced housing, town houses and apartment buildings has been encouraged by the Auckland Regional Growth Strategy (1999). This trend promotes a more compact urban form, through capital investment in various planned and ad hoc redevelopment projects (especially in central areas), and by widespread subdivision of larger suburban properties by landowners ('infill housing'). These trends, set against a backdrop of steady population growth, have been linked to marked increases in vehicular congestion, and associated phenomena such as on-street parking – a known risk factor for child pedestrian injury (Lawson and Edwards 1991). Critics frequently contend that the region's already inadequate roading infrastructure is being utterly overwhelmed by the move towards medium- and high-density housing (Dupuis and Dixon 2002). Notably, this transition has not been accompanied by significant shifts towards greater use of public transport and active travel modes; indeed, car ownership rates continue to rise, in both relative and absolute terms. Aucklanders own almost one car for every two people, and car ownership is growing at about twice the rate of the population. According to the Auckland Regional Council (ARC), nearly 64 per cent of employees drive to work (ARC 2005). At the same time, greenfield sites on the urban fringe are not immune from developments similar to the master-planned communities observed in various Australian, US and Canadian cities.

The growing volumes of vehicles on Auckland roads (and the public attention given them, which frequently loses any sense of international perspective) arguably exacerbates parental fears for children's well-being on city streets, and contribute to the long-term shift towards chauffeuring. This shift cannot be dismissed as simply irrational, for pedestrian injury represents a leading cause of childhood morbidity and mortality in Auckland (see Collins and Kearns 2005). It is part of a broad retreat from the public realm – simultaneously normative and physical – that encompasses efforts to insulate children from the varied dangers perceived to be lurking beyond the family home. In wealthier suburbs, this often takes the form of dividing children's leisure time between the semi-fortified space of the home, and organized activities accessed by private motor vehicle.

Such privatization of children's life worlds is not necessarily accessible to those who live in devalorized urban landscapes – although the imperative of protecting children from 'public dangers' is undoubtedly more pressing in these places. Indeed, in the worst instances they may be 'sites of desertion' where children can know only the 'jaded and foul' (Gleeson 2004: 16–17). Attention to such environments is critical, not least because of their influence on children's well-being and life chances. In an earlier paper (Collins and Kearns 2005), we noted significant international evidence of a steep socio-economic gradient in child injury rates. For example, a major study from Norwich, England, found that injuries were 1.35 times more likely in the most deprived neighbourhoods than in the least deprived, *after* variations in family characteristics were accounted for (Haynes, Reading and Gale 2003). Recent research in Vancouver, Canada, has shown that neighbourhood social economic status (SES) also has an independent effect on children's weight. After controlling for individual age and gender, and family income and education, Oliver and Hayes (2005) found the odds of a child living in a low SES neighbourhood being overweight to be 1.29 times that of a child living in a high SES neighbourhood. Notably, care-givers in low SES neighbourhoods were three times more likely to perceive a lack of safe parks and playspaces in their local area than those in high SES neighbourhoods. Such studies establish that less affluent neighbourhoods have social and physical characteristics that reduce children's opportunities for healthy lives, but also indicate that the neighbourhood may be an appropriate site for interventions intended to promote population health.

In the Auckland context, WSBs are generally perceived positively by children, as they offer enjoyable opportunities for exercise and socializing. As one child involved in a previous study put it, 'I love the talking, the exercise . . . my favourite things are talking and telling jokes' (cited in Kearns, Collins and Beuwelt 2003: 289). These same benefits are also identified by parents, for whom the WSB offers freedom from the drudgery of the daily (motorized) school run (Kearns and Collins 2003a). As one respondent in the present study put it, instead of 'driving and fighting for a car park', parents can 'hop, skip and jump, literally, and . . . have a lot of fun'. Moreover, one WSB coordinator contacted in earlier research noted that WSBs may facilitate the types of social mixing that critical literature suggests are increasingly scarce within polarizing and privatizing cities:

> Children get to know other children, who they see in the playground but don't know their names. There's a mixing of ages, not just people from their own class so it's really good for them . . . Children get to know other adults, who they can trust and talk to, like if they are on the street and they are in trouble or something . . .
>
> (cited in Collins and Kearns 2005: 65)

In low- and high-income neighbourhoods alike, WSBs have the potential to get children out of cars, on their feet and interacting with others, thus contributing to physical well-being, environmental awareness and social cohesion.

Although one recent British study found that children who walked to school were no more active over the course of a week than those who were chauffeured (Metcalf *et al.* 2004), the WSB surely represents a useful step in broader efforts to address inactivity and overweight in children. In signalling such benefits, we are nonetheless critical of some aspects of the WSB phenomenon – it is dependent upon adult supervision and control of children, and may readily represent another form of discipline in their lives. Involvement in a WSB, as either adult supervisor or child passenger, typically entails self-regulation and the embodiment of a responsible, rule-abiding person. In some instances, schools have insisted on regulating bus timetables as well as the behaviour of participants, imposing 'even greater discipline than is . . . evident in (normally) compliant children and uniformed volunteers walking in unison' (Kearns and Collins 2003b: 208). While participation in a WSB is one form of agency in the contemporary city, it can simultaneously require conformity.

In addition, WSB initiatives do not signal a return to relatively unstructured and unsupervised walking or cycling on the part of children. Rather, they contribute to the normalization of constant adult surveillance of children in public space. Strict controls imposed on their operation – most in the name of road safety and risk reduction – can be interpreted as partial acquiescence to the dominance of the motor vehicle. Ultimately, WSBs regulate children rather than traffic (Kearns, Collins and Neuwelt 2003). Such issues notwithstanding, in this chapter we contend that WSBs have the potential to address particular problems associated with intensifying cities.

Methods

The present research was conducted in late 2003, slightly over three years after the first WSB was established in the Auckland region. It sought to survey all schools in the Auckland region that operated WSBs in order to estimate the numbers of children involved, calculate the car journeys saved, identify the benefits and challenges encountered, and gain insight into the future viability of WSB schemes. The respondents were WSB coordinators – the individuals, normally parents, responsible for day-to-day management of each school's WSB route(s).

A list of the 55 schools in the Auckland region with WSBs – complete with coordinator contact details – was obtained from the Auckland Regional Council. The survey was conducted by way of telephone interviews, which lasted 25–40 minutes. In total, 44 interviews were conducted, representing 80 per cent of all schools at which WSB routes were thought to operate. However, two interview transcripts were subsequently excluded, as the initiative had ceased to function at the schools in question. At 27 of the 42 schools for which data was obtained, more than one WSB route was in operation. For the sake of accuracy, respondents were asked about the route they were most actively involved in (e.g.

as a volunteer 'driver' or 'conductor'), rather than the group of routes operating at their school.

Auckland's WSBs

The WSB initiative continued to grow in Auckland in 2003: the number of primary schools operating at least one route increased to 53, from 34 the previous year, and the total number of routes in the region is estimated to have grown to 106, from 63 in 2002. Throughout the Auckland region, 17 per cent of all primary schools now operate a WSB (53 out of 311). Of the 42 coordinators we interviewed, 24 (57 per cent) were associated with schools that had begun WSBs in the previous 12 months. Our survey data is therefore weighted towards recent adopters of the innovation. Fifteen of our respondents reported that only one route operated at their school, while 17 reported two routes, and one school had a remarkable six routes.

Earlier research undertaken in Auckland highlighted the uneven distribution of WSBs throughout the region in 2002, and their concentration in areas of socio-economic privilege (Collins and Kearns 2005). Figure 7.3 shows that this trend persisted in 2003, with the distribution of WSBs by school decile ratings – a marker of socio-economic status – continuing to follow a steep gradient in favour of more affluent school communities.[1]

On an average day, 677 children were participating in the WSBs across the 42 schools for which we were able to collect data. Projecting this figure to

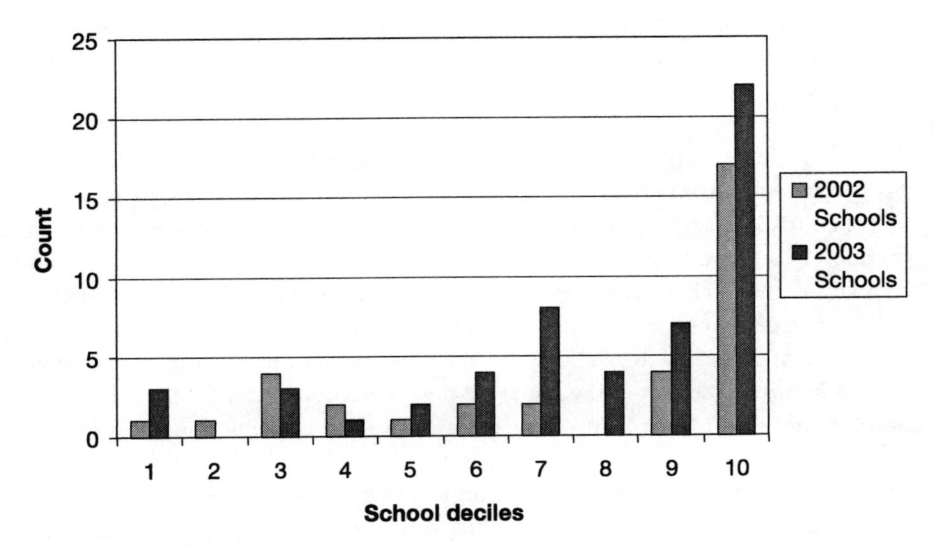

7.3 Number of Auckland walking school buses by decile, 2002 and 2003.

Source: authors.

the 106 routes presumed to be operating at the time of the survey, we can estimate that there were 1,738 children walking on an average day. However, on rainy days the number of children walking fell by approximately 40 per cent, and some routes did not operate at all. The 42 parent coordinators were also asked to estimate the number of car journeys saved by the WSB scheme at their school. Extrapolating the total yielded to all 53 schools with WSB routes in 2003, we calculate that 5,234 car trips were being saved per week.

Attractions, benefits and challenges of WSBs

A substantial proportion (25 per cent) of representatives of the schools that had recently adopted the scheme identified the ability of WSBs to address traffic congestion as the key reason for their establishment. This finding suggests a behaviour change on the part of parents that not only implicates children's activity patterns but also addresses a negative externality associated with population growth and urban intensification. Child health (16 per cent) and safety (13 per cent) benefits were less frequently cited, signalling perhaps the dominance of traffic congestion as a public concern, especially in central Auckland. Some respondents mentioned multiple benefits of the WSB. According to one parent coordinator:

> I instigated it because I realized the absurdity of driving my daughter to school 1.5 km every day. I was embarrassed at the wastage of energy – cars, mental, time etc. . . . I always walked or cycled to school and I realized [driving] was a very inefficient way to get her to school . . . There are great benefits – lots of intangible benefits, over and above the obvious benefits. It knits together the social fabric of your immediate neighbourhood. They get to know other kids' parents. And it's great for the parents as well . . .

The key challenge experienced by the parent coordinators centred on the issue of recruiting and maintaining the support of parent 'drivers' and 'conductors' (51 per cent of responses). WSBs are organized by dedicated volunteers who often struggle to staff the initiative they are committed to. Other concerns centre on the recruitment of child 'passengers' (21 per cent), the efforts involved in organization (13 per cent), and obtaining school support for the scheme (9 per cent).

We asked coordinators about the extent to which they felt that WSBs had resulted in a reduction in traffic congestion at the school gate. Of the 42 parent coordinators, six (14 per cent) reported a high level of reduction, 10 (24 per cent) a modest reduction and five (12 per cent) a slight reduction. One coordinator who had observed a slight reduction noted that the school roll had grown significantly, but at the same time walking was increasingly popular, and there were 'far more independent walkers than before'. In a city with increasing population density, such roll growth is not uncommon. Perhaps surprisingly, a total of 15 respondents (38 per cent) believed the WSBs had resulted in no reduction in congestion at the school gate. In these cases we assume that traffic congestion is sufficiently dense and chronic to appear unaffected by the popularity of the WSB. Developments beyond school gates can have a significant effect on local traffic levels. Indeed, one coordinator commented specifically on the increased levels of traffic associated with the consolidation of housing: 'Now there's a high-density housing development being built across the road from the school – how can we reduce the congestion when Council does this?' Interestingly, in research that asked Auckland primary school children to identify hazards in their neighbourhoods, construction activities associated with gentrification were mentioned with some frequency, alongside on-street parking and busy intersections (Mitchell 2005).

Participation

There is mounting empirical evidence of the health-related benefits of social cohesion as a by-product of participation in community activities (e.g. Baum *et al.* 2000; Ellaway, Macintyre and Kearns 2001). In the present study, the number of parents involved in any one WSB route ranged from only one (at a poorer 'decile one' school) to 20 (a more affluent 'decile eight' school). This suggestion of a gradient in parental participation mirrors the broader gradient in the socio-spatial development of WSBs discussed earlier. However, at two of the schools in poorer areas, other relatives were regular helpers: '[We have] three uncles and two grandmas. One has diabetes so has to walk and this helps her motivation as well as involving her in the community'; 'He's a granddad and he does all the afternoons and all the children will not go anywhere without him. . . . And if he's sick or can't make it, I'm the only one they'll go with.'

Clearly, WSBs are potentially activities invoking physical as well as social benefits for a range of people associated with the school community. This said, most (66 per cent) of the sample reported no contribution by teachers to the day-to-day operation of the WSB, and a minority of schools expressed a 'hands-off' ethos, and a determination to leave the initiative solely in the hands of parents: '[The] principal thinks it will reflect badly on our school if there are any problems and won't have the school associated with it.' One such potential problem is, of course, the risk that one or more children walking on a bus will be struck by a motor vehicle. In some cases, this is a very real fear:

> Council won't put a [stop] light on the crossing on [a nearby road]. Four kids have been hit there. Once a motorbike was going so fast and wasn't going to stop but I made him and he had to skid then fell off, and used all this abusive language to me in front of the kids. It's just not good enough. . . . We – my ladies and me – do a barrier across the road and the kids walk between us. So if anyone gets hit it's us first.

Such comments suggest a high degree of selflessness among some WSB volunteers. They also serve as a vivid reminder of the very real road safety risks that have contributed to children's retreat from the street, which WSBs have only a partial ability to address.

Perception of key benefits

While our previous research offered anecdotal insights into the benefits of WSBs (see Kearns, Collins and Neuwelt 2003), we used the opportunity of this survey to assess their relative significance in the eyes of coordinators. To this end, respondents were presented with a list of benefits identified in the earlier work and asked to identify the first, second and third most important. These preferences were then attributed weights of 3, 2 and 1 'points' respectively, enabling benefits to be ranked according to their perceived importance. To explore whether perceptions varied according to the socio-economic status of school communities, the views of coordinators from samples of high and low decile schools were also tabulated separately. The results of this analysis are presented in Table 7.1. To facilitate comparisons between the three tabulations, the weighted scores are also expressed in relative terms, as proportions of the total number of 'points' available in each.

This data indicates that in the overall sample, children's exercise was the strongest perceived benefit, signalling a broad awareness of the health-promoting elements of walking, and perhaps also of the ways in which alternative opportunities for children's active travel are being curtailed in the intensifying, privatizing city. Following closely was the alleviation of traffic congestion – an issue which, as mentioned previously, has a high profile in the Auckland region, as well as acute local manifestations. The relative prominence of stranger danger was somewhat surprising, as this issue has very seldom been mentioned in response to more free-form questions (e.g. when 30 respondents were asked why their WSB had been established, not one mentioned stranger danger, whereas 19 mentioned congestion). It appears that coordinators see WSBs as offering children protection against strangers when prompted to think about this issue, but that otherwise stranger dangers are not at the forefront of their thinking.

Table 7.1 also allows us to compare two groups of six schools, grouped by socio-economic status. Although the sub-samples are small, the results indicate that the key benefit for the poorer school communities appears to be reducing stranger danger, while exercise for children is the pre-eminent benefit perceived by the more affluent group. The former finding might be interpreted as indicating that the poorer school communities in Auckland are broadly perceived as unsafe relative to their more affluent counterparts, an observation recently corroborated in research with children themselves (Mitchell 2005).

Table 7.1 Perceived benefits of walking school buses

Benefit	*All respondents (n = 42)*		*Low decile sample (n = 6)*		*High decile sample (n = 6)*	
	A	*B*	*A*	*B*	*A*	*B*
Exercise for children	80	0.32	6	0.17	14	0.39
Alleviate traffic congestion	67	0.27	4	0.10	8	0.22
Reduce stranger danger	63	0.25	15	0.42	4	0.10
Injury prevention	19	0.08	5	0.14	6	0.17
Save parents' time	13	0.05	2	0.06	3	0.08
Safety from bullying	7	0.03	4	0.10	0	0.00
Safety from dogs	3	0.01	0	0.00	1	0.03
Totals	252	1.01	36	0.99	36	0.99

Col. A = Weighted score; Col. B = Relative importance.
Source: authors.

Sustainability of walking school buses

According to the predictions of regional planners, the pressure on Auckland's urban space is set to continue. Given current growth scenarios, the long-term viability of interventions such as the WSB is critical. How sustainable is the WSB programme in Auckland? We propose three indicators: their longevity, growth and support (Kearns and Collins 2003a). As most in our sample had only very recently adopted the initiative, our consideration of these indicators is necessarily tentative.

First, we speculate that the longer a volunteer involvement continues, the longer it is likely to last. After a 'honeymoon' period of enthusiasm and novelty, a commitment such as participation in the WSB can become anchored in the routines of not only particular families but also the life of the school. In total, 27 of the surveyed schools had been operating WSBs for over a year. This is an increase from the mere three schools that had this lifespan in our 2002 study (Kearns and Collins 2003a). Conversely, only two schools have completely abandoned the WSB idea.

Growth can be considered a second indicator of the sustainability of WSBs in Auckland. The presence of a growing number of school communities with WSBs can serve to 'normalize' this activity and ensure its legitimacy as both a form of mobility for children and as a form of volunteer commitment for parents. In addition to the considerable growth in the numbers of WSB routes in Auckland, we also note the continued perception on the part of coordinators that most WSB routes have either stable, or increasing, patronage.

A third indicator of programme sustainability is the degree to which WSB initiatives are supported by parents and schools. We calculated the level of parental support as, on average, one parent rostered to help at least once per week for every 1.9 children regularly walking the bus. This level of parental support, together with support from schools (e.g. teachers involved in coordination) and public agencies (e.g. establishment grants and training offered by local government), cumulatively form a context within which WSBs are likely to continue to thrive.

Of arguably more long-term importance is the potential contribution of WSBs to environmental sustainability. These initiatives operate, in part, to reduce families' dependence on the private car for short, neighbourhood-level journeys, and their potential to reduce congestion (and by extension, fuel consumption) is highly valued in Auckland, by both participants (see Table 7.1) and policy makers. By interrupting the vicious cycle of increased car dependence on the school run and making children accustomed to active travel, WSBs offer interconnecting opportunities for collective and personal well-being. The decision by one family to drive their children to school exposes other children to increased risk, prompting their parents to respond in kind (Collins and Kearns 2001). Indeed, the link between low-impact forms of mobility and cities that are simultaneously child friendly and sustainable merits an explicit place in the various materials used to promote WSB initiatives.

Reflecting on the rapid rise and apparent strength of the WSB phenomenon in Auckland, we can identify several 'lessons' for other cities. First, *ongoing evaluation* of the initiative has been critical to its success. An annual commitment on the part of the Auckland Regional Transport Authority to fund a 'stocktake' has provided crucial evidence to gain further leverage in the policy environment. For instance, the identification of an average 19.5 car trips saved per day as a result of the first WSB in Auckland convinced a regional transport agency to make one-off payments of NZ$1,500 to schools adopting the scheme. Subsequent monitoring has provided an impetus for further funding from government bodies, and for regular commentaries in public forums (especially the print media).

Second, *incentives for children* have been important in increasing and maintaining support. Many schools have developed schemes to reward children for their participation. At one school, for instance, walkers receive a 'ticket', which is clipped on each journey, and this record of participation is converted into 'house points' at the end of each term. Children are also issued with certificates detailing the distance they have travelled. While these seemed to be minor additions to the scheme, such recognitions have helped children to own the initiative.

Third, WSB activities have yielded *neighbourhood improvements*, which illustrate the symbolic clout that institutionalized walking may bring. What can otherwise be a marginalized activity within the intensifying city (Solnit 2000) is galvanized through its association with primary schools, its routine operation and its official sanction by various bodies of local and regional government. Upgraded neighbourhood infrastructure due to the operation of a WSB was reported by 27 (64 per cent) of the coordinators surveyed, with the most common improvements relating to footpaths, including the trimming of overhanging trees, maintenance of broken pavement and the removal of obstacles. While successfully lobbying for clearer footpaths may seem a minor victory, it is surely a step towards achieving more significant interventions in favour of child friendly cities. Notably, New Zealand currently lacks any system of low-speed zones on roads near the places where children gather – such as parks and schools (around which 30-kilometre-per-hour speed limits exist in parts of Canada, for example).

Conclusions

At a general level, proposals to address child pedestrian safety concerns have tended to focus on either agency-based responses, such as educating the young about road safety, or structurally based proposals to alter the built environment (e.g. installing speed bumps and other traffic-calming devices). There are inherent limitations associated with both these perspectives. Educating children about road safety can only be effective if children are permitted to be walking and experiencing the neighbourhood environment. Attempting to teach road safety through simulated instruction, while in practice children continue to be chauffeured, is destined to have limited effect. Similarly, educating children and encouraging independent mobility within congested areas, simply invites a victim-blaming response when, eventually, a child injury or fatality occurs. At the other extreme, those who advocate environmental change as the solution commendably critique the hegemony of the motor vehicle and invoke the rights of children as road users. However, the likelihood of truly de-prioritizing motor vehicles in our cities remains slim. The pro-car lobby maintains a firm grip on most city planning processes, including Auckland's (Mees and Dodson 2002).

While WSBs are not without shortcomings, they offer the significant benefit of addressing safety within the car-dominated city, from both educational and environmental perspectives. First, through the act of supervised walking, children can learn in a real-world setting, but be guided in the process. Second, by walking as a group, children reclaim the local environment and become a routine presence on the street. This alters the perceptual environment for motorists. It serves as a reminder that there are other and more vulnerable road users in a safer and more effective way than solitary pedestrians could achieve. As discussed, the WSB can also alter the actual environment, through becoming leverage with which to achieve minor, but significant, improvements in infrastructure.

Within intensifying cities such as Auckland, there is a growing sense that development is fast approaching the limits of comfort in terms of liveability. Significantly, intensification is occurring in the absence of publicly understood planning philosophies, and is only implicitly justified. The success and popularity of WSB schemes provides an example of people responding in support of children in the intensifying city. Their popularity in Auckland shows no signs of abating, notwithstanding the challenges identified by our respondents. However, we would be remiss if we regarded them as a panacea in promoting child pedestrian safety, traffic decongestion and child friendly cities more generally. WSBs have developed a specific purpose and role in the daily rhythms of residential neighbourhoods in Auckland. Because of this specificity, they cannot be relied upon to address a number of other broader public health challenges, such as effecting positive change in driver behaviours, or, more generally, shifting the priorities of the urban political system towards the needs of children.

In this chapter, we have implicitly considered a child friendly city as a healthy community, an urban place founded on ideas of well-being that transcend negative views of health as a 'state of absences'. Within this place, individuals and groups contribute to the common wealth, and in so doing secure themselves a positive place in the world that extends beyond sites of privatized consumption (Kearns, McCreanor and Witten 2006). In this respect, the uneven geography of WSBs in Auckland – their popularity remains largely a middle-class phenomenon – is concerning, pointing as it does to another manifestation of the omnipresent social gradient in health status and resource access. Yet, this said, there is also a danger in reflexively associating conditions of material deprivation with ill-health, social disconnection and despondency. Such responses may encourage defeatism and institutional abandonment of poorer areas.

Notwithstanding empirical evidence about the health-related benefits of the social cohesion stemming from community participation (e.g. Ellaway, Macintyre and Kearns 2001), there is a growing body of literature reporting a decline in a sense of community, based on neighbourhood and physically proximate ties (Kearns, McCreanor and Witten 2006). We have argued that WSBs in Auckland offer a starting point for re-invigorating social cohesion, or, as one respondent eloquently put it, 'knitting the neighbourhood together'. It is the mundane venues of daily life, such as the streets, that ultimately support or inhibit good health and a sense of freedom and belonging (Kearns McCreanor and Witten 2006). As a planning philosophy, urban intensification and related ideas of more compact cities hold promise of safer, healthier settlements (Williams, Burton and Jenks 1996). However the Achilles' heel of these ideas is the private motor vehicle. Given ongoing reluctance to break automobile dependence and promote and use public transport networks, children are destined to remain marginalized users of urban space. As Davis and Jones (1996: 108) have contended:

> healthy children [are] those who are able to access and use city streets for work and play, move about their local area with a reasonable degree of independence and safety,

play some part in local decision-making and have some sense of ownership or entitlement to be heard.

The WSB is a street-level intervention that promises positive steps in all these dimensions.

Note

1 All New Zealand state schools are ranked according to the social, economic and ethnic backgrounds of their pupils. Decile 1 schools are the 10 per cent of schools with the highest proportion of pupils drawn from areas of low socio-economic status, while decile 10 schools are the 10 per cent of schools with the fewest such pupils. A proportion of government funding is linked to this classification system, with decile 1 schools receiving the most assistance and decile 10 schools the least.

References

Auckland Regional Council (ARC) (2005) *Regional Profile.* Available online: www.arc.govt.nz/arc/about-arc/your-council/regional-profile.cfm (accessed 12 June 2005).

Auckland Regional Growth Strategy (1999) *Regional Growth Forum*, Auckland: Auckland Regional Council.

Baum, F., Bush, R., Modra, C., Murray, C., Cox, E., Alexander, K. and Potter, R. (2000) 'Epidemiology of participation: an Australian community study', *Journal of Epidemiology & Community Health*, 54 (6): 414–23.

Collins, D. and Kearns, R. (2001) 'The safe journeys of an enterprising school: negotiating landscapes of opportunity and risk', *Health and Place,* 7 (4): 293–306.

Collins, D. and Kearns, R. (2005) 'Geographies of inequality: child pedestrian injury and walking school buses in Auckland, New Zealand', *Social Science & Medicine*, 60 (1): 61–9.

Davis, A. and Jones, L. (1996) 'Children in the urban environment: an issue for the new public health agenda', *Health and Place*, 2 (2): 107–13.

Dupuis, A. and Dixon, J. (2002) 'Intensification in Auckland: issues and policy implications', *Urban Policy & Research*, 20: 415–28.

Ellaway, A., Macintyre, S. and Kearns, A. (2001) 'Perceptions of place and health in socially contrasting neighbourhoods', *Urban Studies*, 38(12): 2299–316.

Gleeson, B. (2004) *The Future of Australia's Cities: Making Space for Hope*, Professorial Lecture, Brisbane: Griffith University.

Haynes, R., Reading, R. and Gale, S. (2003) 'Household and neighbourhood risks for injury to 5–14-year-old children', *Social Science & Medicine*, 57: 625–36.

Kearns, R. and Collins, D. (2003a) *An Assessment of Walking School Buses in the Auckland Region*, report, Auckland: Auckland Regional Council.

Kearns, R. and Collins, D. (2003b) 'Crossing roads, crossing boundaries: autonomy, authority and risk in a child pedestrian safety initiative', *Space and Polity,* 7: 193–202.

Kearns, R., Collins. D. and Neuwelt, P. (2003) 'The walking school bus: extending children's geographies?', *Area*, 35 (3): 285–92.

Kearns, R. A., McCreanor, T. N. and Witten, K. (2006) 'Health communities', in Freeman, C. and Thompson-Fawcett, M. (eds) *Living Together*, Oxford: Oxford University Press.

Lawson, S. D. and Edwards, P. J. (1991) 'The involvement of ethnic minorities in road accidents: data from three studies of young pedestrian casualties', *Traffic Engineering Control*, 32: 12–19.

Lyne, M. and Moore, R. (2004) *The Potential Health Impacts of Residential Intensification in Auckland City*, Auckland: School of Population Health, University of Auckland, and School of Applied Sciences, Auckland University of Technology.

Mees, P. and Dodson, J. (2002) 'The American heresy: half a century of transport planning in Auckland', in Holland, P., Stephenson, F. and Wearing, A. (eds) *Geography – Spatial Odyssey. Proceedings of the Third Joint Conference of the New Zealand Geographical Society and the Institute of Australian Geographers*, Dunedin: University of Otago.

Metcalf, B., Voss, L., Jeffery, A., Perkins, J. and Wilkin, T. (2004) *Physical Activity Cost of the School Run: Impact on School Children of Being Driven to School*, EarlyBird 22, *BMJ*, 329: 832–3.

Mitchell, H. (2005) *Through the Children's Eyes: (Re)interpreting the Freedom and Use of Public Space from Children's Perspectives*, MSc thesis, Auckland: School of Geography and Environmental Science, University of Auckland.

O'Brien, C. (2001) *Ontario Walkability Study. Trip to School: Children's Experiences and Aspirations*, York Centre for Applied Sustainability, Ontario.

Oliver, L. and Hayes, M. (2005) 'Neighbourhood socioeconomic status and the prevalence of overweight Canadian children and youth', *Canadian Journal of Public Health*, 95 (6): 415–20.

Solnit, R. (2000) *Wanderlust: A History of Walking*, New York: Penguin.

Tranter, P. and Pawson, E. (2001) 'Children's access to local environments: a case study of Christchurch, New Zealand', *Local Environment*, 6: 27–48.

Troy, P. N. (1996) 'The family and urban policy', in Jenks, M., Burton, E. and Williams, K. (eds) *The Compact City: A Sustainable Urban Form?*, New York: Routledge.

Williams, K., Burton, E. and Jenks, M. (1996) 'Achieving the compact city through intensification: an acceptable option?', in Jenks, M., Burton, E. and Williams, K. (eds) *The Compact City: A Sustainable Urban Form?*, New York: Routledge.

Chapter 8

Overcoming social traps

A key to creating child friendly cities

Paul Tranter

Introduction

The notion of entrapment, an issue rarely discussed in urban literature, may provide a key to the creation of child friendly cities. Many parents are caught in 'social traps' that affect decisions about their children's freedom. For example, how does a parent decide whether to let their own child walk or cycle to certain places, especially to school, if there is uncertainty about what other parents are deciding to let their children do? Parents facing such a dilemma feel unable to let their children walk or cycle because of dangers created by other parents who take their children by car. While such decisions are made in the best interest of individual children, the collective impact of these decisions can be that children's freedom to explore their own neighbourhood or city is reduced.

Children's freedom of movement, along with having a diversity of environmental resources to facilitate play, has long been recognized as being of fundamental importance in a child friendly environment (Hart 1979; Moore 1986). The opportunity to move freely within their own neighbourhood is recognized by children themselves as one of the main positive indicators of an urban environment (Chawla 2002). Yet Australian cities may well be in crisis in terms of this indicator of child friendly cities.

The aims of this chapter are to explain the importance of children's freedom to explore and play in their local environment, and to examine strategies for overcoming the social traps that can contribute to declining levels of children's freedom. It focuses on children in middle childhood (8–12 years old). This is the period during which parents start to give their children many 'licences' of independent mobility, such as the licence to walk to school without an adult. While the chapter focuses on children in middle childhood, many of the arguments are likely to have relevance to both younger children and to adolescents.

After outlining the impacts of depriving children of freedom to explore and play in their local environment, the chapter provides international comparisons to demonstrate the low levels of freedom of children in Australian cities. It then outlines possible reasons for declining levels of children's freedom in Australian cities. One of these is the growing strength of social traps that many parents become caught in. The chapter outlines strategies that may allow parents to escape such social traps, hence providing more child friendly Australian cities. It concludes with an explanation of the importance of a multi-faceted approach to the creation of child friendly cities.

Impacts of depriving children of freedom

Most parents are aware of risks their children are exposed to if they are given more freedom to walk or cycle to school and to other places in their neighbourhood or city. They may be less aware of the possible risks of *not* letting their children have this freedom. These risks concern children's physical health and development, opportunities for spontaneous play, exposure to pollutants and their loss of a sense of place.

There has been considerable discussion in the media in Australia (and elsewhere) in recent years about children's lack of physical exercise and the associated high rates of childhood obesity. Overweight and obesity affect about a quarter of children and adolescents in Australia. '[T]he prevalence of overweight children almost doubled, and the prevalence of obese children more than tripled' between 1985 and 1995 (Waters and Baur 2003). Linked to childhood obesity is an increasing rate of Type 2 diabetes and heart and liver problems.

As well as physical health, there is also a risk that children's cognitive, social and emotional development may suffer if children do not have the stimulation that comes from close physical contact with their surroundings (Kytta 2004). Several studies have found that playing in nature has positive impacts on 'children's social play, concentration and motor ability' (Fjortoft and Sageie 2000: 84). Children's social skills and emotional development may be enhanced through having to deal with real world situations in the course of their independent movement through their neighbourhoods (Kegerreis 1993).

If children do not have freedom to explore their environment, they miss opportunities for spontaneous play. Play is widely regarded as being supremely important for children. Children's play also needs to be allowed to happen, rather than to be taught by an adult. There is an important distinction between play and adult directed intellectual, cultural, sporting and leisure activities. These can provide important stimulation for the child, but are not a substitute for (child-directed) play (Tranter and Doyle 1996). Even the journey to and from school can be a playful experience for children if they are allowed to walk or cycle. Many adults can remember the joy and carefree experiences that they had on the way home from school – jumping in puddles, popping tar bubbles, collecting stones or autumn leaves, watching water run down drains, and sharing stories with their friends.

Another impact of depriving children of freedom relates to increased levels of pollution. If parents decide to 'protect' their children by driving them to school, sport, recreation and their friends, they are also contributing to a raised level of pollution (and traffic danger) for other city residents. They are exposing their own children to 'in-car pollution', which 'may pose one of the greatest modern threats to human health' (International Center for Technology Assessment 2000: 5). In-car pollution can be much higher than pollution levels experienced by pedestrians and cyclists (Rank, Folke and Jespersen 2001). 'Elevated in-car pollution concentrations particularly endanger children, the elderly, and people with asthma and other respiratory conditions' (International Center for Technology Assessment 2000: 5).

Perhaps the most insidious impact of depriving children of their freedom to explore their own neighbourhood is the impact it has on children's sense of place. Children's understanding of place is best developed from sensory-rich experiences (Orr 1992) in which children have direct contact with both the natural aspects of their environment and the people in their local community, and ideally have the opportunity to creatively shape their own places (see Figure 8.1). David Engwicht has highlighted the importance of this sense of place for children: 'freedom to explore the local neighbourhood . . . gives [children] an opportunity to develop a relationship with the placeness of their physical environment. Robbing children of a sense of place robs them of the very essence of life' (Engwicht 1992: 39).

Children's freedom in Australian cities

Available evidence indicates that the level of children's freedoms in Australian (and New Zealand) cities is lower than in many other countries, and that these levels of freedom have been declining over recent decades, and may still be in a state of decline (Corpuz and Hay 2005; Tranter 1996; Tranter and Malone 2003).

Though most Australian children do not suffer the extremes of disadvantage that many children suffer in other parts of the world (e.g. inadequate water supply or housing), many Australian children have less freedom to explore their local neighbourhood or city than children living in poverty. For example, in Braybrook (suburban Melbourne), where young people are constructed as a problem and hence removed from the streets, they are less able to participate in community life than, for example, the disadvantaged young people of the very low income Boca-Baraccas area in Buenos Aires (Chawla 2002).

In comparison with children in German cities, children in Australian cities have lower levels of freedom to walk to school alone, cycle on main roads alone, visit friends alone, use public transport, cross main roads alone and go out after dark. Data collected in the 1990s show that while there is considerable variation in children's freedom within cities, children in Australian and New Zealand cities

8.1 Children's sense of place is enhanced by close contact with their physical environment and the opportunity to manipulate their environment: here children play in their own 'store' in a forest.

Source: Paul Tranter.

have much lower levels of freedom than children in German cities (Hillman, Adams, and Whitelegg 1990; Tranter 1996). For example, while 80 per cent of ten-year-old children in German cities were allowed to travel to places other than school alone in 1990, only 37 per cent of ten-year-old children in the Australasian schools surveyed were given this licence in 1992 (see Figure 8.2).

The independent mobility of Australian children has declined significantly over the last 10 to 30 years. Data for Adelaide on modes of transport to and from school show a dramatic decline in numbers of children walking and cycling to school between 1981 and 1997 (see Table 8.1). Data on the journey to school for the Essendon area of Melbourne show an even more pronounced shift to less child friendly transport modes between 1974 and 2005 (Table 8.2).

For the greater Sydney metropolitan area, there have been significant changes in children's independent travel to school in the ten years from 1991 to 2001. During this period, the percentage of children cycling to school (both primary and secondary) halved (from 1.8 per cent to 0.9 per cent), while the percentage taken by motor vehicle increased from 37 per cent to 49 per cent (Cadzow 2004). The Sydney Household Travel Survey found that during the period August 1998 to July 1999, the number of walking trips on weekdays for children younger than 16 years fell by 5 per cent, while trips as car passengers rose by the same proportion (Newton 2001).

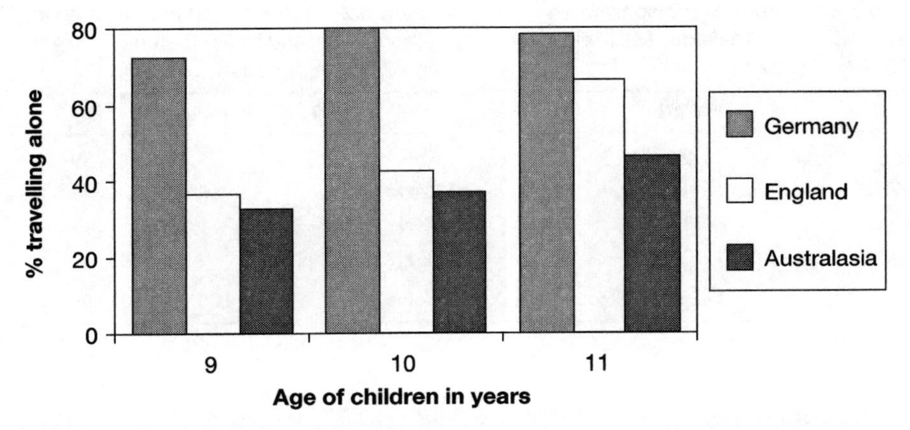

8.2 Percentage of children allowed to travel alone to places other than school for cities in Germany, England and Australasia.

Source: Tranter 1996.

Why is children's freedom so low in Australian cities?

The reasons for the low and declining levels of children's freedoms in Australia are complex and interrelated. Two of the most important reasons are the high (and growing) levels of traffic, and increasing 'stranger danger' fears. There are, however, a number of other issues that impact on children's freedoms.

The design of many modern Australian housing estates, with less space for outdoor activities and more internal space, encourages sedentary indoor activities for children (Gleeson 2004). More houses have a street face dominated by double or triple garages, with automatic roller doors. The amount of house frontage needed for such garaging means that there are fewer windows facing onto the street, giving less of a feeling that the street is being 'watched over'. There are also fewer adults at home during the day, therefore fewer people around who know the neighbourhood children and can keep an eye out for them.

Parental perceptions of the role of parks and playgrounds may be another factor in children's loss of freedom. There is a widespread belief among Australian parents that children's needs for play are met by parks and playgrounds (Tranter 1993). However, reliance on parks and playgrounds is a poor substitute for a truly child friendly city (Cunningham and Jones 1999), as suggested in the following quote, '[o]ne should be able to play everywhere, easily, loosely, and not forced into a "playground" or "park". The failure of an urban environment can be measured in direct proportion to the number of "playgrounds"' (Ward 1977: 73).

Children's freedom has also been eroded through the impact of competing demands on their time. Sometimes this is imposed by the parents, who insist on children being involved in various extra-curricular activities (music, sport,

Table 8.1 Modes of transport to and from school in Adelaide, 1981 and 1997 (in per cent)

	1981	*1997*
Car	24	60
Walk	42	20.5
Cycle	14	4.5
Other transport	20	15
Total	100	100

Source: Bikewest 2001.

Table 8.2 Modes of transport to primary school in Essendon (Melbourne), 1974 and 2005 (in per cent)

	1974	*2005*
Car	25	89
Walk	65	8
Cycle	1	1
Public transport	9	2
Total	100	100

Source: Peddie and Somerville 2005.

ballet, maths coaching, etc.). It is also child-driven, particularly in terms of the popularity of the internet, computer games and television. For many children, their only experience of the outside world is a virtual experience in a simulation game. But, as Gleeson argues in Chapter 3, 'a tour of duty in SimCity can never emulate the sensuous complexity of urban life. Life with the Sims is unlikely to help a child cope with any family dysfunctionality or prepare them for the joys and strains of adult life'.

'The empirical evidence on the reasons why so many children are taken to school by car suggests that it is to do with the complexity of modern life and the resulting shortage of time' (Mackett 2002: 32). This complexity is at least in part due to the impact of the car. Most people believe that cars save them time. But the 'effective speed' of cars is often slower than a bus or a bicycle (Whitelegg 1993a, 1993b; Tranter 2004). This is because when we think of the time that a car saves us, we do not consider the substantial amount of time we are obliged to spend at work to earn the money to pay all the costs associated with car use.

Ironically, another reason for the restricted freedom of Australian children is the fixation that many parents have with providing the 'best' possible life chances for their children. Parents who are obsessed with providing their children with opportunities for succeeding in a consumerist and materialist world sometimes unwittingly deprive their own children (and other children) of freedom to explore and play. Their own children's lives become over-scheduled, over-occupied and over-organized (Chawla 2002) to the point where there is no time for them to experience the 'joy and wonder' that comes with having the time to explore their local neighbourhood (O'Brien 2003). Other children's opportunities are reduced through the effect of the extra traffic generated by parents driving their children long distances to 'the best schools', 'the best child-care centres' or 'the best maths tutors' (Morris, Wang and Lilja 2001).

Not only are many children being driven to schools and other places that are further from home than closer (if not 'the best') alternatives, but the last few decades have seen an increasing dispersion of facilities and schools, partly due to the dominance of private motor vehicles. Government policy dominated by neoliberal ideology has led to the closure of many local schools and post offices as well as children's services and health facilities (Gleeson 2004). Consequently, many destinations are now seen as too distant to walk or cycle to.

The final reason to be discussed for the low level of children's freedoms in Australia relates to the concept of 'social traps' or 'social dilemmas' (Hallsworth, Black and Tolley 1995; Black 1997; Kitamura, Nakayama and Yamamoto 1999).

Social traps (or social dilemmas)

The concept of social traps applies very clearly to the freedoms that parents give their children to walk or cycle to school (and to other places). There are two components to this social trap in terms of the journey to school. First, many parents feel 'trapped' into driving their children because the majority of other parents are contributing to a dangerous situation by driving their children to school. Second, parents may fear social alienation if they do not conform to the social pressure to be a 'good parent'.

Many parents believe that cars are being used unnecessarily for the journey to school, but use the extra traffic around schools as a reason for driving their own child (Tranter and Pawson 2001) (see Figure 8.3). The most dangerous part of many children's journey home from school is contending with the traffic jams created by parents driving to school to pick up their children to protect them from traffic danger. In Victoria, 65 per cent of child pedestrian accidents in the age group 4–13 years occurred just before or after school (Morris, Wang and Lilja 2001).

If all parents allowed their children to walk or cycle to school, they would be safer in terms of both traffic danger and stranger danger. They would be safer from traffic, as there would be fewer cars (especially near schools). They would also be safer from strangers because they would be in larger groups, and hence could benefit from 'safety in numbers'. However, if only one parent allows their child to walk/cycle, that child will suffer the negative consequences of other people's actions. As the social trap engages, fewer and fewer children (and adults) are left walking and cycling. Those that do continue to walk or cycle are increasingly endangered.

In some areas of cities where traffic levels are low and where children have access to a wide range of natural or 'wild' spaces, lack of freedom for children is more likely to be a function of the particular culture of 'being a good parent' that may have developed within a community. This implies strong social pressures that are relevant to 'social traps'. Parents may feel that they are 'expected' to drive their children because that is what all the other parents are doing, and allowing their children to walk or cycle, or play on the street may be seen as irresponsible parenting (Tranter and Pawson 2001).

8.3 Parents dropping children at a Canberra primary school: many parents believe that cars are being used unnecessarily for the journey to school, but use the extra traffic around schools as a reason for driving their own child to school.

Source: Paul Tranter.

How can social traps be overcome?

Social traps operate most effectively when there is no opportunity for collective decision-making: no forum for communication between individuals who are making independent decisions (e.g. about their children's travel). Most people poorly understand the collective impact of many individual decisions, and even those that do understand it often feel powerless to reduce this impact.

Before parents can escape from social traps they need to be able to:

- understand the collective impacts of their individual decisions;
- come to a collective agreement to change their behaviour in ways that will benefit all of their children (and themselves).

Dr Catherine O'Brien has conducted some innovative action research on the travel behaviours of Canadian children and their parents. In one project, she presented to parents a summary of the negative impacts of cars on children (most of which are the collective impacts of individual travel decisions) (O'Brien 2001). She tapped into the innate desire of parents to protect their children from dangers and to provide children with a stimulating and enjoyable childhood. Parents were shocked to discover the extent of the negative impacts of cars on children.

O'Brien found that when parents are told about these impacts, including those from their own cars, they become more open to changing their behaviour to decrease their car use. The things that concerned them most were 'the loss of spontaneous play opportunities and restricted independent mobility' (O'Brien 2001: 7).

It is not enough to expect large numbers of individual parents to change their behaviour, just because they are aware of the negative impacts of their own travel behaviour. The need for a collective agreement must be addressed.

Overcoming social traps with travel behaviour change programmes

In Australia there is a range of programmes aimed at voluntary travel behaviour change, many of which use some form of the 'TravelSmart' brand name. These programmes aim to reduce the total number of car trips by empowering people to walk, cycle and use public transport, and to build community support for such changes (James 2002: 41). One component of travel behaviour change strategies in Australia involves state-based programmes linked with voluntary travel behaviour change in schools (e.g. TravelSmart to School in Western Australia, TravelSmart School programme in Queensland, Brisbane City Council's Active School Travel programme and TravelSmart Education – Schools in Victoria). Such programmes are referred to generically here as 'Travel Behaviour Change Schools' programmes.

The Travel Behaviour Change Schools programmes currently being implemented in several Australian states provide a potentially effective way of overcoming the social traps discussed above. Travel Behaviour Change School programmes are particularly important in Australian society because of the lack of local neighbourhood-based communities in many Australian suburbs (Doeksen 1997; Morris, Wang and Lilja 2001). 'The cult of the individual has become the dominant social ethos' (Cunningham 1996: 89). Without some sense of local community, it is almost impossible to break out of social traps, as 'the individual' takes precedence over any altruistic community-based behaviours. Schools provide one of the few foci for the strengthening of this locally based community, and can be used to help parents move from an individualist strategy for enhancing their children's safety and well-being to a community based strategy. Travel Behaviour Change Schools programmes tap into the bond between members of the school community, and in a process of positive feedback, they can strengthen these bonds and further develop the school community.

The strategies required to overcome social traps are reflected in the terminology used in Travel Behaviour Change Schools programmes. For example, the objectives of Travel Behaviour Change Schools programmes in Victoria include:

> [to] enhance community building as families work together to plan and share responsibility for children using more active ways of getting to and from school and around the local travel environment . . . to engage all members of the school community – school councils, administrators, teachers, students, parents/carers and other family members . . . Broader school communities may wish to adopt a TravelSmart program . . . to partially resolve a common problem of many primary schools – the confusion and danger at drop off and pick up times in and around schools . . . [The programmes aim to] create social pressure to change ways to travel.
>
> (Dynamic Outcomes 2004)

Policy makers working in Travel Behaviour Change (Peddie 2004, personal communication; Twine 2004, personal communication) appreciate the usefulness of the concept of social traps, and can see how their Travel Behaviour Change programmes are likely to help undo such traps.

Travel Behaviour Change Schools programmes depend for their success on achieving a 'critical mass' of parents who are willing to embrace new ideas and not be constrained by individualistic thinking. The crucial difference between current Travel Behaviour Change Schools programmes in Australian cities and some previous initiatives of local councils in the UK, for example, is that many schools programmes in Australia concentrate on this critical mass of parents in individual schools rather than on a series of ad hoc strategies (e.g. 'walking school bus', or 'how to catch a bus'). While such programmes are intrinsically worthwhile, they have limited impact on behaviour change by themselves (Kearns, Collins and Neuwelt 2003). Travel Behaviour Change programmes can facilitate communication among a critical mass of parents (along with teachers, children and other members of the local community).

Within the school community, ongoing communication is needed first, to get parents to appreciate the need for change and the collective benefit of such change. The next stage involves encouraging parents not to regress to their previous driving behaviours. Eventually, not only can the social trap be released but social pressure can develop to conform to new, enlightened views, such as letting children walk to school. Of course, this may need to be supported by initiatives such as the walking school bus, particularly for younger children.

The success of the Travel Behaviour Change programmes in overcoming social traps is dependent on many factors, but one essential factor is the commitment and enthusiasm of the school travel coordinator (or equivalent position). This person can be either a parent or a teacher, who works within the school community to raise awareness and generate the critical mass of parent (and teacher) support. An interesting research project would involve examining the size of the critical mass that is needed in order to break the social traps.

The various schools programmes in Australia have been boosted by the publicity given to physical activity guidelines and the recognition of high levels of childhood obesity in Australia. Travel Behaviour Change Schools programmes have been able to promote travel behaviour change by highlighting the real potential of 'active transport' (walking, cycling and public transport) to address these problems.

School community meetings show that there is a widespread awareness of the problems of car-based school journeys, and a groundswell of public support for addressing issues such as health, social exclusion (the impact of isolating children in metal cocoons), the importance of getting more people on the street (walking, cycling, socially interacting and playing), and the financial rewards to parents (e.g. if they can avoid purchasing the second car). However, planning meetings are still faced with the problem of parents thinking as individuals.

In a reference group meeting for a schools programme in a Brisbane school, one parent took an individualistic view on the issue of car-pooling. This parent could recognize the need to address the issue of large distance (too far to walk or cycle) and the problems associated with 'a 4WD backing out of every house in the morning'. Yet, despite being able to recognize the collective negative impact of this situation, she was reluctant to car pool because she 'didn't trust the other drivers'. At this meeting a local police representative showed some underlying awareness of social traps, when he encouraged parents by suggesting: 'you've got to take a broader view; you have got to think of the broader good' (Twine 2004, personal communication).

Independent evaluations of TravelSmart programmes have questioned some claims about their effectiveness (Stopher and Bullock 2003). In their critique of TravelSmart programmes, Stopher and Bullock (2003) argue that the location chosen for the implementation of these programmes may be biased towards areas with greater potential for change to environmentally friendly modes of transport, yet they are still recognized as having a positive impact on travel behaviour change.

There have been other criticisms of aspects of Travel Behaviour Change Schools programmes. Initiatives such as the Walking School Bus initiative have been criticized, for example by Mayer Hillman, a British researcher who has investigated children's freedom in England and Germany. According to Hillman, walking school buses, though introduced with the best of intentions:

> promote paranoia among parents that they are not acting responsibly unless they are always with their children outside the home and . . . make them feel irresponsible if they are unable to do so or to delegate someone to act in their place.
>
> (Hillman 1999)

Despite these criticisms, schools (and children) provide an important focus in travel behaviour change. One of the advantages of a focus on schools is that children themselves can be involved in finding ways to increase levels of walking and cycling. Children may have creative ideas and innovative solutions to travel behaviour change. This also has long-term effects on sustainable transport. If we can change the mind-set of children so that they see an alternative to car travel, at least for short intra-urban trips, then when they become adults they may be less inclined to turn immediately to the car: '[B]ehaviours adopted in childhood are often carried through into adulthood' (Di Pietro and Hughes 2003). Unfortunately, many Australian children become conditioned to believe that the car is the best mode of transport, even for short trips.

School travel is not the only travel by children that involves car travel, but it is easier to target through travel behaviour change programmes (Morris, Wang and Lilja 2001). This is because it allows programmes to work with a clearly defined community (parents, teachers and children), the members of which can more easily appreciate the benefits of working together. Also, trips by car to school are a significant proportion of total travel – 'trips accompanying children to school' represent about 20 per cent of the total trips made in Melbourne in the 8.30 a.m. to 9.00 a.m. peak (Morris, Wang and Lilja 2001).

Travel to School programmes are also an ideal way to promote regular exercise. While Australian government policy concentrates on encouraging (and glorifying) sporting activity, this may be misguided. 'Issues of motivation, cost and availability' (Morris, Wang and Lilja 2001), as well as the extra traffic generated by weekend and after-school sport, mean that the total impact of sporting programmes on physical fitness and health is likely to be less than the impact of encouraging more walking and cycling to school – activities that provide regular 'active transport'.

Even if trips to school account for a minority of children's overall travel activity (Hillman 1999), encouraging them to walk or cycle to school may well translate into giving them more confidence in walking and cycling elsewhere. Their parents may also be more accepting of this if children can demonstrate their competence in walking or cycling to school. 'Once children have started to use the alternatives to the car, and realize the benefits this gives them in terms of independence, they may start to use alternatives for other trips' (Mackett 2002: 37).

Conclusion

The arguments in this chapter demonstrate that Australian cities do not rate highly on one key indicator of child friendly cities: levels of freedom for children to explore and play in their own neighbourhood or city. Of more concern is the likelihood that, without significant planning and policy changes, levels of independent mobility of Australian children may decline further over the next decade. Ironically, part of the reason for the continuing decline in children's freedom is that many parents are trying to give children the best opportunities, and trying to protect them from dangers (particularly traffic danger and stranger danger). However, parents are not considering the collective impact of their individual decisions. While creating child friendly cities will require a multi-faceted approach, Travel Behaviour Change Schools programmes may help parents to consider such collective impacts, and to find a way out of the social traps that lead to high levels of car use for children's transport. Children may once again be able to experience a sense of joy and wonder, even for their journeys to and from school.

Social traps are difficult to escape when motorized traffic levels are high and when there are low levels of local neighbourhood-based community interactions. These traps are strengthened by both a culture of increasing use of the car as well as the changing urban structure and form of Australian cities. One of the fastest growing types of urban travel involves chauffeuring trips, many of which are made to protect children from the very traffic that these trips contribute to. Levels of motorized traffic are increased when urban consolidation (or intensification) applies only to residential development (Chapter 7), and when government policy leads to a greater dispersal of shops, post offices, health care services and schools.

The implications of this for policy are that the achievement of child friendly cities will require a combination of widespread implementation of Travel Behaviour Change programmes, along with a fundamental change in government policy that affects the distribution of services. Planning the location of schools and other services should be done with the needs of children (and non-motorists) in mind, rather than with the narrow objective of reducing the cost of service provision of government departments. Such child friendly planning is likely to provide benefits for all city residents, not just for children. When cities are designed so that children can safely explore their local environment, their parents have more time and more money, the environment benefits (from less pollution and traffic congestion), and the whole community benefits from having fitter, happier and healthier neighbourhoods.

References

Bikewest (2001) *Cycle helmets*, available online: www.cycle-helmets.com/bikewest.html (accessed 18 October 2003).

Black, C. S. (1997) *Behavioural Dimensions of the Transport Sustainability Problem*, Ph.D. thesis, University of Portsmouth.

Cadzow, J. (2004) 'The bubble-wrap generation', *Sydney Morning Herald Good Weekend Magazine*, 17 January: 18–21.

Chawla, L. (2002) *Growing Up in an Urbanizing World*, London: Earthscan/UNESCO.

Corpuz, G. and Hay, A. (2005) 'Walking for transport and health: trends in Sydney in the last decade', 28th Australasian Transport Research Forum, *Transporting the Future: Transport in a Changing Environment*, 28–30 September 2005, Sofitel Wentworth Hotel, Sydney.

Cunningham, C. (1996) 'A philosophical framework for urban planning: the concept of altruistic surplus', in van der Meulen, G. G. and Erkelens, P. A. (eds) *Urban Habitat: The Environment of Tomorrow,* Delft: Eindhoven University of Technology.

Cunningham, C. and Jones, M. A. (1999) 'The playground: a confession of failure?', *Built Environment,* 25 (1): 11–17.

Di Pietro, G. and Hughes, I. (2003) *TravelSmart Schools – There Really is a Better Way to Go,* paper presented at the 24th Australasian Transport Research Forum, Wellington.

Doeksen, H. (1997) 'Reducing crime and the fear of crime by reclaiming New Zealand's suburban street', *Landscape and Urban Planning,* 39 (2): 243–52.

Dynamic Outcomes (2004) *TravelSmart Schools: Better Ways to Go: Methodology Report,* Melbourne: Victorian Government.

Engwicht, D. (1992) *Towards an Eco -City: Calming the Traffic.* Sydney: Envirobook.

Fjortoft, I. and Sageie, J. (2000) 'The natural environment as a playground for children: Landscape description and analyses of a natural landscape', *Landscape and Urban Planning,* 48 (1–2): 83–97.

Gleeson, B. (2004) *The Future of Australia's Cities: Making Space for Hope,* Professorial Lecture, Brisbane: Griffith University.

Hallsworth, A. G., Black, C. S. and Tolley, R. (1995) 'Psycho-social dimensions of public quiescence towards risks from traffic generated atmospheric pollution', *Journal of Transport Geography,* 3 (1): 39–51.

Hart, R. (1979) *Children's Experience of Place.* New York: Irvington.

Hillman, M. (1999) *The Impact of Transport Policy on Children's Development,* Canterbury safe routes to schools project seminar, Canterbury Christ Church University College, 29 May, available online: www.spokeseastkent.org.uk/mayer.htm (accessed 5 May 2004).

Hillman, M., Adams, J. and Whitelegg, J. (1990) *One False Move: A Study of Children's Independent Mobility,* London: Policy Studies Institute.

International Center for Technology Assessment (2000) *In-car Air Pollution: The Hidden Threat to Automobile Drivers, Report No. 4, An Assessment of the Air Quality Inside Automobile Passenger Compartments,* available online: www.icta.org/doc/In-car%20 pollution%20report.pdf (accessed 15 September 2004).

James, B. (2002) 'TravelSmart – large-scale cost-effective mobility management: experiences from Perth, Western Australia', *Municipal Engineer,* 15 (1): 39–48.

Kearns, R. A., Collins, D. C. A. and Neuwelt, P. M. (2003) 'The walking school bus: extending children's geographies?', *Area,* 35 (3): 285–92.

Kegerreis, S. (1993) 'Independent mobility and children's mental and emotional development', in Hillman, M. (ed.) *Children, Transport and the Quality of Life,* London: PSI Press.

Kitamura, R., Nakayama, S. and Yamamoto, T. (1999) 'Self-reinforcing motorization: can travel demand management take us out of the social trap?', *Transport Policy,* 6: 135–45.

Kytta, M. (2004) 'The extent of children's independent mobility and the number of actualized affordances as criteria for child friendly environments', *Journal of Environmental Psychology,* 24: 179–98.

Mackett, R. L. (2002) 'Increasing car dependency of children: should we be worried?', *Municipal Engineer,* 151 (1): 29–38.

Moore, R. C. (1986) *Childhood's Domain: Play and Place in Child Development,* London: Croom Helm.

Morris, J., Wang, F. and Lilja, L. (2001) 'School children's travel patterns – a look back and a way forward', paper presented at the 24th Australasian Transport Research Forum: *Zero Road Toll – A Dream or a Realistic Vision,* Hobart.

Newton, P. W. (2001) *Human Settlements Theme Report: Liveability – Human Well-being – Transport Demand, Access and Congestion*, available online: www.deh.gov.au/soe/2001/settlements/settlements03-3a.html (accessed 15 October 2003).

O'Brien, C. (2001) 'Children: a critical link for changing driving behaviour', *National Center for Bicycling and Walking (NCBW) Forum (Canada)*, 52: 4–13.

O'Brien, C. (2003) 'Transportation that's actually good for the soul', *National Center for Bicycling and Walking (NCBW) Forum (Canada)*, 54: 1–13.

Orr, D. (1992) *Ecological Literacy: Education and the Transition to a Postmodern World*, New York: State University of New York Press.

Peddie, B. (2004) Personal communication, Project Manager, TravelSmart Education, Department of Infrastructure, Melbourne, Victoria.

Peddie, B. and Somerville, C. (2005) 'The ghost of TOD's past: schools reconnecting', Transit Oriented Development Conference, 2–8 July, Fremantle.

Rank, J., Folke, J. and Jespersen, P. H. (2001) 'Differences in cyclists and car drivers exposure to air pollution from traffic in the city of Copenhagen', *The Science of the Total Environment*, 279: 131–6.

Stopher, P. and Bullock, P. (2003) *TravelSmart: A Critical Appraisal*, Institute of Transport Studies Working Paper 03 14, Sydney: The University of Sydney.

Tranter, P. J. (1993) *Children's Mobility in Canberra: Confinement or Independence?*, Monograph Series No. 7, Canberra: Department of Geography and Oceanography, University College, Australian Defence Force Academy.

Tranter, P. J. (1996) 'Children's independent mobility and urban form in Australasian, English and German cities', in Hensher, D., King, J. and Oum, T. (eds) *World Transport Research: Proceedings of the Seventh World Conference on Transport Research, Volume 3: Transport Policy*, Sydney: World Conference on Transport Research.

Tranter, P. J. (2004) *Effective Speeds: Car Costs Are Slowing Us Down*, Department of the Environment and Heritage, Australian Greenhouse Office, available online: www.greenhouse.gov.au/tdm/publications/pubs/effectivespeeds.pdf (accessed 3 March 2005).

Tranter, P. J. and Doyle, J. (1996) 'Reclaiming the residential street as playspace', *International Play Journal*, 4: 81–97.

Tranter, P. J. and Malone, K. (2003) *Out of Bounds: Insights from Children to Support a Cultural Shift Towards Sustainable and Child-friendly Cities*, paper presented at the State of Australian Cities National Conference, Parramatta.

Tranter, P. J. and Pawson, E. (2001) 'Children's access to local environments: a case-study of Christchurch, New Zealand', *Local Environment*, 6 (1): 27–48.

Twine, A. (2004) Personal communication, Principal Active Transport, Transport and Traffic, Brisbane Administration Centre, Queensland.

Ward, C. (1977) *The Child in the City*. London: Architectural Press.

Waters, E. B. and Baur, L. A. (2003) 'Childhood obesity: modernity's scourge', *Medical Journal of Australia*, 178 (9): 422–3.

Whitelegg, J. (1993a) 'Time pollution', *The Ecologist*, 23 (4): 131–4.

Whitelegg, J. (1993b) *Transport for a Sustainable Future: the Case for Europe*. London: Belhaven Press.

Chapter 9

Creating child friendly playspaces

A practitioner's perspective[1]

Prue Walsh

Introduction

Previous chapters in this volume have described how various socio-cultural and environmental changes in modern Western societies, particularly in Australia and New Zealand, are impacting on the experiences of children and young people living in cities. These changes include the contraction of the public domain, parental colonization of children's life worlds (Chapter 3) and intensification of the city (Chapter 7). In many contemporary cities, these changes have forced children indoors and away from the public realm, with harmful potential consequences for their development. It is in the public realm that children learn many essential life skills, including how to socialize, make decisions, solve problems and gain a sense of belonging.

As well as being able to access the public domain, children should also have the opportunity to play spontaneously and freely in the urban environment and not be forced into playgrounds or parks (Ward 1977). This suggests a fundamental re-examination of the influences on a child's development processes, especially those operating in the first five years of their lives (Shonkoff and Phillips 2000). When urban design decisions do not consider the importance of play, society is adversely affected. Lynch (2004) found that for every US$1 invested in high-quality early childhood development programmes, an economic return of US$13 is gained.

To be effective, play should heighten children's awareness, capture their interest and act as an ongoing motivational tool to explore the world in which they live. It is through this process that continuous learning occurs. The effect of play is profound because it stimulates and enhances child development. While play can occur anywhere, important play opportunities can be provided outside the home. Where else can a child:

- explore the natural environment – contributing to cognitive development;
- move with speed – improving gross motor development; and
- share robust play with groups of children – adding to social development?

At the same time, however, it should be noted that playgrounds and parks can significantly contribute to the play opportunities provided to children in cities, where other forms of public realm, notably open space, are often difficult to access. Public playspaces are also social assets of the community, providing a place where adults can meet while their children play, and where senior citizens can observe this play and feel part of the wider community. The problem with children's playspace design tends to derive not from the deficiencies of playgrounds – though these can be real and of concern – but more usually from the lack of the additional recreational spaces that encourage spontaneous free play. This points to the need for a more critical understanding of how the layout and design of suburbs, not simply playgrounds, affects children's well-being.

This chapter focuses on how urban playspaces can better meet the needs of children and the community. The focus is on the contemporary Australian urban experience, but the analysis draws from an international literature. The arguments and suggestions made reflect the insights from a practitioner's perspective, based upon long professional experience of urban playspace development in Australia and in other developed nations. The insights are practical, but, as noted along the way, find confirmation in several of the scholarly assessments made earlier in this volume.

The chapter begins by charting some of the forces that are changing the prospects for children's play in contemporary cities, with an emphasis on the Australian experience. This is followed by a critical discussion of some of the current development practices that affect children's play prospects in cities. Some broad principles that can be applied to the planning of community playspaces will then be identified. The next section presents guidelines for the planning of public playspaces at the subdivision and internal design levels. The chapter finishes by outlining a brief for the design of children's urban and suburban playspaces.

The contemporary urban context for play

There are a number of shifts taking place in contemporary Western societies that are reshaping children's opportunities for play, thus altering their social experiences and, potentially, their developmental abilities. The consequences of these trends will be explored in a later section, which addresses current development practices.

First, the wider process of the contraction of the public domain, which has been recasting the landscapes of many Western cities, is contributing to the inadequacy of public spaces for children. At the centre of the issue are the forces of increasing individualism, diminishing respect for the public sphere, and the growth of economic insecurity, which Gleeson in Chapter 3 identifies as three major urban consequences of neo-liberal reform (see also Chapter 5). At the same time, heightened general fears concerning public liability have been associated with growing risk aversion among local governments. In Australia, one consequence of the rise of the 'municipal risk culture' has been a decline in the quality of public spaces for children. Boring play equipment in playgrounds and parks replicated time and time again is the physical manifestation of this risk culture. Ultimately, children bear its costs.

Second, the intensification of the city (Chapter 7), characterized by increasing population densities, dwellings and often worsening traffic congestion, has significant implications for public and private spaces for children in the city. Many playspaces are intensely used, increasingly rationed (as land is lost to redevelopment), restricted and hazardous to access due to congestion. Problems for children in the intensifying city are most obvious in the inner city, where densities are at a peak. However, the increasingly generalized 'compaction' policy adopted in many contemporary Australian cities in response to perceptions of metropolitan sprawl may ensure that intensification pressures on children spread beyond the inner city to middle and outer ring areas. As noted above, the problem already seems to be manifesting in newer fringe estates characterized by large houses on small lots, and frequently minimal public domain. The lack of diverse, high-quality public playspaces for children that plagues many inner cities may become more evident in outer locations as well.

Third, children's domestic environments in today's society are somewhat different to those of past generations. Yesterday's suburbs tended to be 'one size fits all' landscapes, reflecting broad institutional and development industry assumptions that households were overwhelmingly large nuclear families that needed detached dwellings with large backyards. As children became more independent, play occurred in neighbour's yards, on the street or in public spaces. This notion of 'one size fits all' did not explicitly cater to all children's play needs. In practice, the subdivisions that formed around its assumptions did usually allocate a generous amount of formal and informal playspace to children, especially in Australian and New Zealand suburbia. The problems inherent in suburban playspace have, however, become more pronounced in recent times. In today's suburbs, a child's play environment is quite different. Socially it tends to mean fewer siblings, carers spending longer hours in the workforce, higher parental expectations and a perception of public spaces as unsafe for children. Physically it translates into children being trapped inside built structures such as childcare centres, domestic dwellings and video arcades, or in supervised outdoor activities for a larger portion of time. In addition, the growth of the contemporary suburban 'megahouse' in the newer estates of many Australian cities has often been at the expense of children's playspace, meaning that play increasingly occurs largely within the confines of the home.

The role of community playspaces in a child's development should not be underestimated. As pointed out by Freeman in Chapter 5, it is in the wider public realm where children learn many life skills. This is where they learn about society and their place within it. Public playspaces also provide children with the space to make contact with the natural environment and a meeting point where relationships with other children and adults in the community can be formed.

Orr (1992) has described how it is through these experiences that children develop an understanding of place and identity. Without adequate community playspaces children are deprived of the many developmentally beneficial experiences these spaces offer. The question then must be what kinds of playspaces are being created that do not encourage or invite children to play?

Current planning and design practice

City builders and designers come from a variety of backgrounds: town planning, traffic engineering, water or electrical supply engineering, architecture, etc. The problem for children is that there is no single authority empowered to advise on their playspaces. In many cases, the only guidance for local planning control comes from outdated and poorly researched state regulations. In other cases there is a guideline vacuum, particularly with respect to inner city childcare and shopping centre occasional care. The situation is not much better for schools. In Australia, the Queensland government's facilities brief for new primary schools does not address playground provision except for a small diagram depicting a climbing structure, softfall surface and fixed shade shelter.

While Australia is a signatory to the Rights of the Child convention, it is surprising that its state governments are not moving to develop guidelines that ensure that built environments provide healthy playspaces for children. In the interim, it would be prudent to consult some of the existing best practice guidelines (Walsh 1996). The result is that many urban playspaces are not only inadequate but sometimes antagonistic to children. Children of all ages should be able to access playspaces and use them with enthusiasm and joy – to find spaces that invite, not simply permit, them to play. For play opportunities to be interesting and able to sustain children's interest over long periods of time, they have to be adaptable and capable of meeting age-related skills, expanding interests and developmental levels. Such play facilities are described as 'open-ended' play options, as opposed to 'closed' play options; the latter is descriptive of the fixed metal structures currently pervasive in community playgrounds.

Closed play options result from a failure to understand both how children play and what it is that stimulates and sustains their interest. Instead, the design of closed play equipment has been based on adults' perceptions of children's play, a practice referred to in the literature as 'urban myths' or 'monuments to misunderstanding' (Evans 1987). Worryingly, playspace myths tend to spread pervasively, reflected in the frequent practice of uncritically copying designs from one area and imposing them in many others. These practices show little understanding of how children play, or why one space is inviting and another is boring. Models are unthinkingly perpetuated by professionals and institutions (Featherstone 2002) and yet playspaces should be geared to children's perceptions, not adults' (Walsh 1992).

One of the major problems inherent in mindless emulation is that playspace safety shortfalls can also be copied. This can be self-defeating: a developer who does not appreciate the litigation risks that result from such shortfalls is not prudent. Playground safety standards tend to be extremely technical, hiding many devils in technical detail, and precedents for misunderstanding are well established. However, a playspace that is 'too safe' also tends to be boring. Children actively seek challenge (and thereby develop risk assessment skills), but risk does need to be carefully managed so that major trauma is avoided. A practical balance needs to be found; some precautions (such as softfall surfaces or freefall zones) must always be put in place, but other play opportunities (such as decks and creeks) should also be considered for their open-ended nature.

The plight of children in inner city areas is usually worse than that in the suburbs because land is so expensive and the needs of these 'non-consumers' tend therefore to be undervalued. This problem was largely ignored for decades in Australia and New Zealand as inner city areas lost population, and it was thought that suburbanization would solve all ills. But now inner city areas in both countries are experiencing a surge in high-rise living, and attention needs to be refocused on the needs of downtown kids. The design of playspaces for inner urban children and youths inevitably encounters the land value and land shortage problems, and therefore needs to be based upon lateral, flexible solutions. Without desirable playspaces, there is likely to be a growth in dysfunctional behaviours such as graffiti or vandalism.

There are some lessons to be learnt from international experience regarding the provision of community playspaces in tight urban areas. An example can be found in the roof top gardens of some European cities, illustrating that good design of tight spaces can be achieved. London's Diana Princess of Wales Memorial Playground provides another good example of inner urban community playspaces that sustain children's interest through a variety of open-ended play options. However, because high density inner city living has not been a major feature of Australian and New Zealand cities for many decades, governments do not have sufficient child focused urban policies or design guidelines in place for such areas. Therefore they often allow redevelopment to occur in ways that leave children out of the equation. Built environments change slowly, meaning that these design errors greatly compromise the possibilities for any future design solutions for children's diverse and evolving needs.

The urban compaction agenda, now well entrenched in many government policies, is also, in some areas, contributing to the segregation of children from the natural environment. One such example exists in Queensland, Australia, where recent reform to childcare legislation[2] allows for the provision of an 'indoor space' in place of an outdoor play area 'in cases where suitable outdoor play areas are not available' (Queensland Government 2004a: 2). Thus, the prioritizing of increased densities takes precedence over providing quality outdoor environments for children's play in inner city childcare centres.

It is purported that the policy responds to an increasing demand for the provision of childcare in the inner city and CBD (Queensland Government 2004b: 1), by removing the precondition of outdoor space and thus encouraging the supply of facilities. Are the facilities that the policy is likely to produce worth the effort? Will they be good for children? On both counts, the policy is doubtful. An additional concern, however, is that the policy amendment will pave the way for developers of childcare centres in outer locations to reduce costs and exclude space for outdoor play. It is politically hard to contain policies that liberalize development conditions for one sector of a city.

In Australia, the rapid growth in childcare centres that are run for profit (Chapter 3) and that aim relentlessly to reduce costs indicates the dangers inherent in any otherwise well-intentioned approach that allows the relaxation of standards. Policies underpinning the amended childcare legislation in Queensland that place low priority on the provision of quality outdoor environments in childcare facilities contribute to the separation of children from their natural surroundings. This separation is proven to have deleterious developmental consequences for children (Orr 1992).

Finally, various studies of conventional suburbs with modest houses on large lots tend to suggest that only 60–80 per cent of a child's independent play will be undertaken at the domestic dwelling or at an institutional venue, such as a childcare centre or school (Cunningham, Jones and Barlow 1996). However, such statistics may have limited relevance to the contemporary planning context, which is marked by the loss of possibilities for independent play generally. This context is increasingly dominated by higher density development and ever more 'intense' urban lifestyles. It reflects rising parental expectations of children and increasing adult concerns about their children's security, in concert with the diminishing ability of many carers to spend time with the young. These environmental and cultural changes may generally be reducing children's prospects for independent play especially in informal playspaces and institutions. Pragmatically, we may need to factor in the growing parental appetite for supervision into playspace planning and design, accepting that it may not be a healthy overall trend for children.

Towards basic planning principles

Planning for children's playspaces has traditionally been a low priority in urban environments, often squeezed into left-over spaces, trouble-prone areas (e.g. poorly drained) or difficult-to-access locations. Too often playspaces are simply tokenistic. The drift away from community provision to for-profit provision, and the rise of cost considerations in service planning, threatens to further marginalize children's needs in cities.

How do we bring children's needs back into the mainstreams of service planning and urban development? Children will play anywhere and at any time, but quality play opportunities can be delivered only through a deliberate process. The following imperatives would be a good starting point for the redesign of playspace parameters:

1. Children have a right to play and their ability or inability to play will impact markedly on their overall competency development. Children are part of the 'residential package' and their unique needs should have special consideration in any new development at every scale, from the individual landholding to the subdivision.
2. As children grow up, their needs change. Therefore an urban area has to service multiple age groups, requiring flexibility/variety of design at all levels.
3. Public playspaces need to be planned and designed to invite child usage and to support the adults who come with them.

These planning principles need to be supplemented in development practice by the design considerations outlined later in this chapter. To ensure these points are recognized in new urban development, there must be changes to the way that children's playspaces are planned for in contemporary cities. A few comments on this urgent task now follow.

Planning guidelines for children's playspaces

Some of the intricacies in general contemporary playspace design for both informal (free, spontaneous play) and formal (organization controlled) spaces include:

1. Dedicated space: play should not be 'fitted in' as an afterthought. It is not possible for teenagers to engage in most ball games without enough space.
2. Access: the opportunity to seek outside venues can be limited to adult-available periods, for example, weekend blocks of time.
3. Variety: children actively seek a variety of play options, for example, street cricket today, sandpit tomorrow, computers or TV another day. Community playspaces are but one of a suite of options deemed desirable by children, which must also include less formal spaces (e.g. bushland reserves) where free, spontaneous play is possible.
4. Invitation: the degree of stimulation achieved by different approaches to community playspace designs needs to be considered. Boring or exciting? Open-ended? Attractive to child and adult users? Sensory rich?
5. Ambience: the climate (or micro-climate) has to be taken into account. For example, does the space include shelter from rain or sun, exposure to

wind (particularly for young children), boggy or dusty places? Are there natural elements present?

The result is a multi-layered solution. In practice there is no simple measurement that can dictate good playspace design. A variety of localized social and climatic considerations must be taken into account in any planning context. At the same time, some general guidelines can be drawn out that are relevant to the different categories of community playspaces.

Community playspaces tend to fall into three categories:

- suburban public parks;
- inner city playspaces; and
- organizationally controlled facilities such as early childhood centres, childcare, schools and sporting venues.

Each playspace form should be considered both as a specific-user venue and, more broadly, as a social facility that should draw from and contribute to the civic resources of the wider community. The 'bottom line' is that if access to them is difficult, they will not be used and they will not integrate with their host community meaningfully (Moore 1990).

Suburban public parks

In contemporary suburban areas, the use of public parks is governed strongly by access. It has been suggested (Moore 1990) that park facilities should be chosen and designed with different layers of access so that they:

- are accessible to people with strollers and the elderly;
- are accessible by public transport and by car; and
- support children's independent access to facilities.

Consider the access routes suggested in the suburban plan shown in Figure 9.1. The points to draw from this design can be summarized as:

- Both car and pedestrian access is convenient to green spaces, parks and school childcare.
- Middle and upper age children have independent access to play areas.
- A choice of local venues is offered, which leads to local 'ownership'.
- Access for maintenance is provided.
- The variety of venues means that micro-design differences are potentially deliverable.
- Cul-de-sac housing also leads to local supervision (Moore and Chawla 1986).

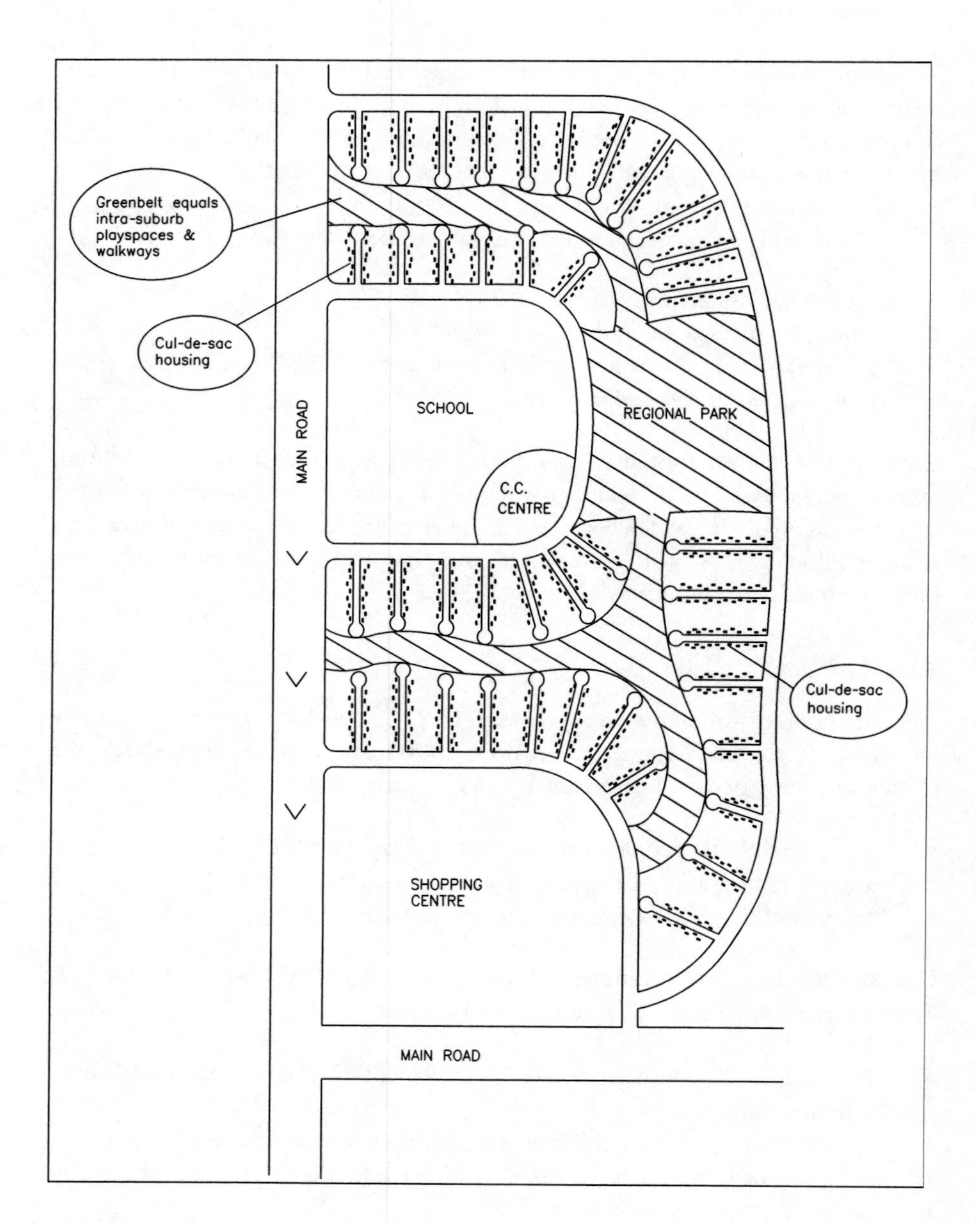

9.1 Suburban plan.

Source: adapted from Moore and Chawla 1986.

Variations of this type of integrated living have been put in place in Australian cities – Canberra is one such city. It is obviously easier to achieve multi-layering of access in new suburban developments than it is in urban re-developments, where the existing built form will constrain many design possibilities.

Some other useful ideas for child focused public park design in new 'green-field' developments are set out in Cunningham, Jones and Barlow (1996). These ideas rest on certain key design values – safety, manipulable spaces (i.e. open-ended), accessibility and user-friendly environments. The main points are:

- There should be natural bushland and a complex manipulable environment within about 200 metres of any house, which should be accessible to children without them having to cross heavily used streets. The landscaping should be complex and designed for a variety of play uses.
- Engineering of residential streets should convey symbolically and practically that 'every road is a bicycle road', as in many European cities.
- The neighbourhood should have clear focal points (schools, pocket parks, sports fields, shops) connected by child friendly streets and/or linear parks with cycleways.

A specific area of planning concern is suburban infill – where there is a loss of natural bushland, sporting facilities and even playspaces connected to suburban shopping centres. The Queensland State Government's new South East Queensland Regional Plan estimates, for example, that about 80 per cent of new dwellings in Brisbane (over the years 2004–16) will be urban infill (Queensland Government 2005). Like many other official documents, this 'blueprint' is long on broad policy visions and short on practical details. Its implementation will have to consider carefully the important details of development and design that are not present in the plan. One key implementation challenge will be attempting to ensure that infill produces child friendly high density environments. Australian redevelopment practice to date has not been encouraging in this respect. Too often, as noted earlier, the super profits on offer during infill redevelopment tend to result in erasure, not the recognition of children's needs in a dense urban environment. These problems are already manifesting in playspaces of many contemporary Australian cities.

Inner city playspaces

Inner city areas tend to be playspace-poor, yet the demand is increasing. Planners need to provide community playspaces in the inner city, accessible to both residents and day users from other urban areas. An increasing number of cities are moving towards plaza (vehicle-free) developments, characterized by trees, water, seating, art works – but no playspaces. A walk through any of these areas will demonstrate that children of shoppers and residents are trying to use and enjoy them, by walking along walls, jumping up stairs, splashing in the fountains. It surely is no great architectural leap to provide simple, robust play opportunities (such as a 1-metre tunnel or a mound to climb and roll down), which the children can enjoy while the parents relax and watch in amenities set aside for them.

A feature of many of the larger contemporary urban shopping centres in both the inner city and the suburbs is the provision of small indoor playspaces provided by centre management and individual stores such as fast food outlets. But these are only available during trading hours, and only provide gross motor activities (equivalent to an adult gym). Play is in fact so much more than the scripted, confined movement dictated in these places.

Controlled facilities

The final category of playspaces is that of organizationally controlled facilities such as childcare centres, schools and sporting facilities. The location and access to such places tends to have a higher planning priority in new suburbs than the informal playspaces discussed above. Most of the problems relating to new facilities lie in site planning and the internal design of the space, which is discussed later.

It is the inner city areas in which formal playspaces are located that there are very real problems and barriers (Walsh 2004). Childcare and schools in these locations tend to consist of blacktop surface outdoor areas, if they are provided at all.[3] Furthermore, buildings are often compromised by ad hoc alterations put in place over many years of operation. The resulting play quality is bad for the children brought for care from outer areas, but is potentially disastrous for the children who live locally in environments that are almost universally 'hardscaped'. In Australia, guidelines to address these issues are being developed through advocacy organizations such as Early Childhood Australia. However, an acceptable solution can only involve a more rigorous set of formal statutory design guidelines, with input from bureaucrats and developers, as well as advocacy organizations.

Design brief for play and play support

From the formal literature and wisdom accumulated during practice, it is possible to distil a design brief for child focused play that incorporates child development principles and child behaviour managing parameters, as well as established standards. This design brief should detail the amount of space needed, how space should be used and the necessary play options and facilities to support play.

The most important of these is to ensure that enough space is allocated for both current use and for future users. Tight space limits the play options, so obviously the more space the better. When children spend long hours in any one locality, they move from one activity to another. Research shows that children need at least four play options available to avoid being crowded by other users (Kritchevsky and Prescott 1977). Berry (1993) found that young children rarely spend more than four minutes on any one fixed climbing frame, once they have mastered the basic skills involved. Children will be attracted to play opportunities that are placed so as to invite a development of ideas – a cramped space cannot provide this.

The amount of space needed is derived from a formula based on the ages of the users, the expected numbers using it at any one time, and the play opportunities provided. Internationally, childcare outdoor space per child ranges from 7 square metres in Canada to 20 square metres in Sweden. Mixed age groups always need more space per capita than homogenous age groups (for example, a family group would need more space per person than a large group such as a Year 5 primary school group). The space required for a small suburban park (not a regional park) to serve 50 mixed age users is in the range of 1,000–1,500 square metres.[4]

The second component of the design brief relates to how that space is used. While many facility designers make the distinction between supervised (childcare, schools) and unsupervised areas, the central fact is that effective play opportunities are those designed to be used within a range of developmentally based skills. The pervasive 'monuments to misunderstanding' (discussed earlier) presented as closed play options, cannot provide for the ongoing challenges of children's play demands. Investment in such facilities provides very little in terms of user satisfaction. They are not very cost-effective – nor do they make for exciting play.

Today there is a better understanding of child development, which results in more age/skill appropriate designs. These are not just scale related but skill related and interest related. They include quiet focused play (gazebo, stage), places that promote imagination and pretend play (elevated platforms, caves, mazes), areas for physically active play such as digging patches, open-ended fixed equipment, open space (ball games), areas for social interaction (seating, hidey holes) and areas to manipulate parts so as to work through an idea (sandpit, creek, nature area). Through all of these the children should be able to feel the thrill of succeeding and to gain the will to keep on trying.

Designing playspaces requires a subtle blend of assessment of terrains and understanding how children play. There are best practice documents commissioned by Australian governments (New South Wales, Northern Territory) and the Lutheran Church (Queensland) that set out some of the Australian design parameters (see Walsh 1996) or overseas design parameters (e.g. Mauffette 1999). Best practice is illustrated in Figure 9.2 for a childcare outdoor play area that can be translated into a stand-alone or shopping-precinct playspace.

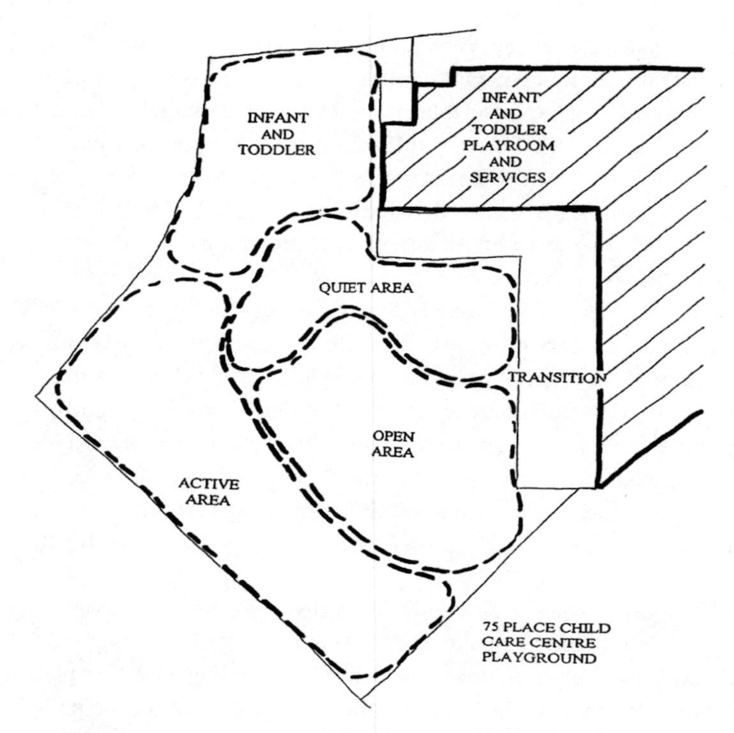

9.2 Best practice for a childcare outdoor play area.
Source: Walsh 1996.

This figure illustrates the following key design features:

- Intrusion is prevented by separating activities by activity type.
- Visual and physical supervision extends throughout the playspace.
- Within each activity-type area, clusters of variable yet compatible play opportunities have been established.
- The playspace is designed with safety and skill level in mind.
- Each user has at least four play options (Kritchevsky and Prescott 1977).

Underpinning all design briefs is the fact that children develop at different rates – and the lack of appropriate play opportunities will adversely affect their development. If the designer does not appreciate how children play, then the facility will be unsatisfactory. Too few play options often result in fights over the use of popular facilities. While children are the initial losers when this occurs, it is the community that ultimately bears the cost.

The third and final component of the design brief is to determine the type of infrastructure needed to support the variety of play options. These considerations tend to be simple and practical, for example:

- Young children need close supervision and support while they play, so supervising adults need adjacent seating with easy visual or physical access to the children, shade, nappy change areas and tables.
- Play options that are inaccessible or too challenging to the developmental level of the individual child can become daunting and the child is unlikely to develop the competency in that skill area (social, physical, cognitive or emotional).
- Older children tend to be more active and get hot/tired, so they need to be able to independently access water fountains and toilets, and to have quiet rest areas between bouts of more active play.
- Teenagers tend to be sociable, so they need group seating (next to large rubbish bins if eating is involved) and an informal ball court.
- Elderly people often like to watch children at play ('covert socializing'), so they need seating, shade and smooth access routes.
- Backup services should be considered, including:
  - perimeter security in the form of fencing or hedges, tailored to the community access norms, and
  - car parking areas and drop-off zones.

Conclusion

This chapter has surveyed some of the shifts occurring in contemporary cities that are undermining the possibilities for healthy play and development for children. The shift to higher densities has often neglected the needs of children. It is unfortunate that planners have pursued this childless development agenda. Further, a variety of social changes such as increased parental surveillance and programming of children threaten to undermine the prospects for urban free play – an activity, indeed a right, that would be cherished in a child friendly city. For the best results in planning, it is necessary to use people with play/child expertise.

The chapter has presented a range of planning principles, guidelines and design specifications to help policy makers and practitioners to better comprehend the diversity of children's play needs. These suggestions are only the beginning and should not constrain design innovation that really has children's interests at heart. Children have unlimited imagination – town planners and developers should have equal imagination, or make way for people who do.

Notes

1 The author gratefully acknowledges the editorial input of Brendan Gleeson and Kylie Rolley from Griffith University, and of Valerie Eldershaw for professional assessment of the concepts underpinning this material.

2 See the Queensland Development Code, Part 22 – amendment effective since November 2004.

3 Specifically in Queensland where the legislation permits childcare centres without outdoor play areas.

4 Calculated as: 50 users × (20 to 30 square metres per user) = 1,000 to 1,500 sq m.

References

Berry, P. (1993) 'Young children's use of fixed playground equipment', *International Play Journal*, 1: 115–31.

Cunningham, C. J., Jones, M. A. and Barlow, M. K. (1996) *Town Planning and Children*, Armidale: Department of Geography and Planning, University of New England.

Evans, J. (1987) 'Playgrounds . . . monuments to misunderstanding', *Child Accident and Prevention Foundation of Australian Conference*, Melbourne, May 1987.

Featherstone, M. (2002) 'The physical environment: some thoughts about design and the design and the design process', *Challenge*, July: 12–19.

Kritchevsky, S. and Prescott, E. (1977) *Planning Environments for Young Children: Physical Space,* 2nd edn, Washington DC: National Association for the Education of Young Children.

Lynch, R. (2004) *Exceptional Returns: Economic, Fiscal and Social Benefits of Investing in Early Childhood Development*, Washington DC: Economic Policy Institute.

Mauffette, A. G. (1999) *Revisiting Children's Outdoor Environments: A Focus on Design, Play and Safety,* Quebec: Gavin Presses.

Moore, G. T. and Chawla, L. (1986) 'Neighbourhood play environments: design principles for latchkey children', *Children's Environments Quarterly* 3 (2): Summer, 13–23.

Moore, R. C. (1990) *Childhood's Domain: Play and Place in Child Development*, Berkeley, CA: MIG Communication.

Orr, D. (1992) *Ecological Literacy: Education and the Transition to a Postmodern World*, New York: State University of New York Press.

Queensland Government (2004a) *Information Paper: Alternative Solution Guidelines for 'Enclosed' Outdoor Play Facilities*, Brisbane: Queensland Government, Department of Communities. Available online: www.communities.qld.gov.au/childcare/resources/documents/pdf/ip_outdoor_play.pdf (accessed 1 August 2005).

Queensland Government (2004b) *Consultation Paper Review Queensland Development Code Part 22 – Child Care Centres*, Brisbane: Queensland Government, Department of Communities. Available online: www.communities.qld.gov.au/childcare/cclegislation/documents/word/consultationpaper.doc (accessed 1 August 2005).

Queensland Government (2005) *South East Queensland Regional Plan 2005–2026*, Brisbane: Queensland Government, Department of Local Government, Sport and Recreation and South East Queensland Regional Organisation of Councils.

Shonkoff, J. and Phillips, D. (eds) (2000) *From Neurons to Neighbourhoods: the Science of Early Childhood Development*, Washington DC: National Academy Press.

Walsh, P. A. (1992) 'Fixed equipment – time for a change', *Australian Journal of Early Childhood*, 17: 3.

Walsh, P. A. (1996) *Best Practice Guidelines in Early Childhood Physical Environments*, Sydney: NSW Department of Community Services, NSW Government Information Services bookshop.

Walsh, P. A. (2004) 'When the quick fix won't do', *Rattler* 72, Summer: 17–20.

Ward, C. (1977) *The Child in the City*, London: Architectural Press.

Chapter 10

Pathways to the child friendly city

Brendan Gleeson, Neil Sipe and Kylie Rolley

We began this volume by stating that the creation of child friendly cities is a comprehensive task that will need to draw from a broad range of specialist knowledge about human development and environmental change. This book captures only some of these understandings, drawing especially from those disciplines and professions that address the human consequences of urban environmental change. From the essays contained within these pages can be glimpsed some of the pathways towards a child friendly city. Other pathways have been well laid out previously – notably, the fundamental need for professional and institutional sensitivity both to children's unique perceptions of built environments and to their particular social and environmental needs in urban settings. Other pathways will emerge from future discussions.

In this brief concluding statement, we have taken the liberty of attempting to distil the main prescriptive content from our contributors' essays. We have used the three-part thematic structure of the book to organize these prescriptive statements into a set of pathways. The 'core sampling' logic behind the structure thus defines the points of departure for these pathways to the child friendly city: from different points of political and spatial scale; from within discrete areas of professional and policy practice; and from within distinct spheres of everyday action. The result is a quilted depiction that captures some of the pathways to the child friendly city. It's not an 'all or nothing' scenario in our view: pursuit of this admittedly partial map will necessarily improve urban life for children. Some of the lessons are so fundamental that any commitment to them by the institutions and professions that most influence urban change will be to the direct benefit of children.

It's wrong to neglect rights

Karen Malone described how at the global scale, nearly half of the world's children are living in poverty, and are denied the basic protections and resources laid out in the CRC. These include the right to safe and clean environments and to spaces for leisure, recreation and free play. These 'rights deficits' also go to the core of broader sustainability concerns. Thus the UN's approach to the issue, and the one advocated by Malone, has been to couple sustainability goals with children's rights as the foundation for a global framework for child friendly cities. The logic behind the notion is that if sustainability goals are not achieved, then children will be the most profoundly affected population group. Therefore the well-being of children can be used as an indicator of sustainability.

Malone reviewed two global initiatives within the UN framework on sustainability and human rights: UNICEF's CFCI; and UNESCO's GUIC project. Two fundamental characteristics of what constitutes a child friendly city were apparent. The first is that the level of governance most suited to the implementation of the CRC is the local level. This is the governance level that has ultimate responsibility for, and the most significant impact on, children's well-being. Second, the welfare of children cannot be perceived by adults acting on behalf of children – instead governments must work towards realizing the potential of children to become 'authentic participants' in decision-making processes. This is an imperative repeatedly identified in other essays in the book. The realization of child friendly cities is dependent upon finding ways for children to become involved in determining their future. This most commonly will mean helping them to contribute to discussions about community problems at the local level.

Material security is important but children need much more

Taking the discussion to the national (Australia) level, Brendan Gleeson provided a critical analysis of the structural shifts in Australian cities that are adversely affecting children from both high- and low-income households. The analysis does not present a promising vision of the future for young Australians if the current directions of political and economic change are allowed to persist. Pursuit of neo-liberal reform has already contributed to: a decline in community-based childhood services and growth in profit-driven care; the proliferation of increasingly lifeless playspaces for children amid mounting concern about public liability; geographical and social polarization that is threatening young lives in low-income households; and increasing incidence of health problems among young people, the most obvious of which is rising levels of obesity.

At the core of these issues are neo-liberal reform and the 'Growth Fetish' it has produced – the result of which is the emergence of toxic cities, not healthy cities. Gleeson's analysis reveals that materialism and individualism are centre-pieces of the neo-liberal ideology that is breaking down the foundations of what constitutes a child friendly city. Hence, it is these forces that must be challenged if child friendly cities are to be achieved. Importantly, the failure of the 'Growth Fetish' reveals the inadequacy of a historically cherished idea that greater household wealth will necessarily improve children's happiness and welfare. A central prescription from this is that far more attention needs to be paid both to the potential costs of affluence for children and to their non-material needs. In the current neo-liberal context, a restoration of parental time and care might be the most pressing need for children.

The cost of collisions: it's cheaper to include children in planning

Claire Freeman reminds us that urban planners play a particularly important role in the creation of child friendly cities. It is planners who have the ability to determine the form and structure of urban environments through the regulation of development trends. They have the power thus to affect significantly the creation of children's urban home worlds. Conventionally, planning decisions are based on what planners believe is in the public interest. Children, however, too often get submerged within a public dominated by adults. The result for urban development scenarios is repeated 'collisions' between adults' and children's worlds, with children coming out the losers more often than not.

To avoid such collisions it is necessary for planners to first acknowledge that children's interests should be explicitly acknowledged and represented in the planning process. Steps have to be taken towards increasing the amount of shared decision-making between adults and children in institutional settings, such as everyday planning practice. For this process to be successful it is necessary for children to gain experience in shared decision-making in other aspects of their lives. Children are very adept at learning new skills and transferring those to unfamiliar contexts. Freeman also exposes the fact that most planners have very little appreciation of children's environmental experiences. Children's experiences are not confined simply to recreational and educational domains: when planning does glimpse children, it is usually only in these settings. Their experiences extend into arenas such as transport, housing, shopping and the interactions between different urban domains. It is essential that planners understand the variety of children's experiences and the implications of planning decisions in each. Adding further complexity to the issue is the compartmentalization of children's interests into discrete levels of government and between separate government agencies and departments. This points to the need for improved governance of children based upon better networks to promote the flow and exchange of information between the governments and the agencies that have responsibility for children.

A rule for the good city: engage the unruly

Kurt Iveson offered a critique of different policy approaches relating to young people in public spaces. He argued that there is a need to move away from neo-liberal frameworks that are based on a vision of the good city as one free of conflict towards a framework that allows divergent interpretations of the city. In pursuit of a city free of conflict, this neo-liberal framework is the foundation of two separate approaches to deal with 'out of control' youth: circuits of *exclusion* and circuits of *inclusion*. Both of these approaches, although seemingly counter-posed, result in the banishment of the anti-social. Both assume that the characteristics of a desirable city are predetermined. The determination rarely involves young people, and never the 'anti-social' young. If cities are to become spaces that are friendly to young people, it has to be acknowledged that there are many interpretations of what makes a good city.

A strategy of *engagement* is Iveson's alternative to the pursuit of exclusion and inclusion. It is an approach that invites varied interpretations of the city. The goal of a policy based on engagement is to develop shared projects where young people can express in their own terms the problems that need to be addressed in the city and how these might be overcome. Underpinning this approach is the notion that discussions about what constitutes a good city should be the fundamental role of urban politics and policy. Children and young people must have the opportunity to participate in these debates.

The journey to integrated knowledge has just begun

Neil Sipe, Nick Buchanan and Jago Dodson reviewed the international literature on children and cities, drawing out the health related themes that have emerged over the past century. It was found that in more recent times children's perceptions of their environment have become a focus of research, as has the study of links between the urban environment and childhood health and obesity. Two key messages emerged from their survey. First, there now exists widespread professional and scientific recognition that pursuit of the child friendly city must be based upon on multidisciplinary knowledge frameworks that directly inform collaborative, cross-sectoral policy interventions. Second, while the interdependencies between the different phenomena that shape children's health and welfare in cities have been recognized, the scientific journey to better understand these links has only just begun. Urban scholars have an important role to play in this enterprise.

The walking school bus, a step not a leap towards the child friendly city

Robin Kearns and Damian Collins addressed the issue of how to create healthy environments for children in the intensifying city. Specifically, the walking school bus (WSB) programme in Auckland was reviewed in terms of its ability to provide a safe alternative to car-dominated travel. The analysis revealed that while the programme was not without its weaknesses, it demonstrated a number of advantages for child participants. First, it provides an opportunity for children to learn the skills needed to negotiate real urban environments while being guided by adults. Second, through continuous experience of the public realm via the WSB, children can achieve an improved connection with their local environmental surroundings. Third, their presence reminds motorists about the existence and the needs of other users of urban space. Fourth, the WSB offers an opportunity for children to become more involved with other members of the community, thereby improving their physical health and feelings of social cohesion. Last, through its ongoing operation, the WSB has potential to generate support for small but significant neighbourhood improvements, such as the maintenance of footpaths and the trimming of overhanging trees.

On the flip side, there are some disadvantages that have been identified in the critique of the WSB. Its operation is reliant upon adult supervision and therefore constitutes an additional form of control in children's already structured lives. Further, the programme can also be seen as a strategy to regulate children in a car-dominated environment, as opposed to challenging the prioritization of the private vehicle over other users. Most concerning though is the revelation that the distribution of WSBs in Auckland (and surely other cities) is uneven, favouring areas of socio-economic privilege, where child health and road safety issues are less urgent. To summarize, the WSB is a beneficial programme that may be a partial solution to the problems of children's physical inactivity and disconnection from the local community and environment. However, it is important that the programme be accessible to children of all socio-economic backgrounds, especially lower income households where need is greatest. A central prescription from these findings is the need for caution when deploying any single 'solution' to an urban problem confronting children. Solutions are usually adult-conceived ideas that might neglect children's real needs. They can also have unintended consequences if conceived naively, and may, for example, worsen social inequality if inequitably applied.

Mind the trap: strategies to overcome the chauffeuring culture

Paul Tranter argued that children's ability to move freely and independently in their environment is a key characteristic of a child friendly environment. Tranter recognized that Australian cities fare very poorly on this score and that it is 'social traps' that are largely responsible for the failing. 'Social traps' occur when parents make decisions about their child's travel behaviour without the knowledge of what other parents might be doing. In this circumstance parents feel trapped both into chauffeuring their child to and from school and into preventing them from cycling or walking. Due to the heavy volumes of traffic around schools, parents then feel it is too dangerous to allow their child to cycle or walk. They also fear they will be perceived as bad parents if they do not conform to the chauffeuring culture.

The key to overcoming social traps is to create opportunities for communication between individuals, especially parents/carers, which can develop into forums for collective decision-making about children's travel. In order to influence parental chauffeuring behaviour, it is first necessary to raise broad awareness of the negative consequences of car-dominated travel on children's health, and then devise an agreement on ways in which travel behaviour can be changed to the benefit of all. Interventions that have addressed the issue of social traps, such as the Australian 'Travel Behaviour Change Schools' programmes, have suggested the following four-step approach is the key to success:

1. Target the school community and try to break through individualistic thinking.
2. Aim for a critical mass of parents.
3. Ensure that communication is ongoing to avoid regressing into previous travel behaviour.
4. Enlist a committed and enthusiastic coordinator.

While educational travel is only a small portion of children's total travel, by eliminating school-based social traps, children's energies and interests can be diverted more generally away from car based journeys towards use of more active travel modes.

Working harder on children's play

Prue Walsh examined playspace design considerations in inner city and suburban environments. Previous chapters (3 and 7) showed that children's lives are increasingly structured and regulated, leaving little time for the free and spontaneous play that is so crucial to children's developmental abilities. In many cases, children have little opportunity beyond the neighbourhood playground, childcare centre or school to access open space and the natural environment. It is imperative, therefore, that the design of these spaces is based on a thorough understanding of children's play needs.

For example, it is fundamental that play facilities are open-ended: that is, adaptable and capable of meeting children's expanding interests and rising developmental competencies. This stands opposed to a design that is based on adults' perceptions of how children play. Second, playspaces have to begin with sufficient space so that redesigns and future improvements are not restricted. Simply designing children's playgrounds to fit into left-over or residual space is not an appropriate way to cater for children's needs. Third, children's playspaces have to be easily accessible to a variety of users, including the elderly and people with strollers, and by foot from all households in the city. Last, playspaces should provide a variety of play options to maintain the interests of children who frequently alternate between different physical activities.

Put simply, neighbourhood playgrounds and formal play facilities should invite children to play and not simply permit them to play. To be successful at making playspaces more inviting, planners, designers and developers must have an understanding of how children play. This is a social necessity: it is through play that children learn many important life skills.

The final word

There is much to be done to create child friendly cities in developed nations, such as Australia and New Zealand, which have been the focus of this book. Karen Malone reminded us of the unique challenges facing developing cities, but also of the universal nature of many of the problems discussed in this book. In this final chapter we have drawn on the work of the contributing authors to present a range of strategies for elevating children's interests in urban affairs generally. Our hope is that the pathways identified will be considered by the professions and institutions that most influence urban conditions. The weight of evidence in this book suggests that children in Western cities urgently need and deserve this sort of attention. The journey to the child friendly city must soon begin in earnest. The destination we seek is not an exclusive wonderland for children. Our destination is a diverse city that places children at its centre because it is committed to universal human values, including care, respect and tolerance. This is no vision of a theme park. It is a vision of human sustainability.

Index

Pages containing illustrations are indicated in *italic* type.